Dearest Cherchie,

1987

I'm praying for you that may your [...] become true. "I know you will follow the desires [...] or when all is old your heart will be the same." [...] day waking up and know that you would be the first woman on Mars. I hope you enjoy this book more than I enjoyed picking it for you. Rememer, a poet once told me,

"I carry your heart, I carry it in my heart"

Have a good Christmas.

Lots of love,
The Mud Species ~
Pankaj

HISTORY OF
NASA

HISTORY OF
NASA

AMERICA'S VOYAGE TO THE STARS

E John and Nancy DeWaard

Exeter Books

NEW YORK

A Bison Book

First published in U.S.A.
by Exeter Books
Distributed by Bookthrift
Exeter is a trademark of Bookthrift Marketing, Inc.
Bookthrift is a registered trademark of Bookthrift Marketing, Inc.
New York, New York

Copyright © 1984 Bison Books Corp.

ISBN 0-671-06983-7

Printed in Hong Kong

Reprinted 1987

Contents

The Dream and the Promise

Somewhere in the vast reaches of the Universe exists a disc-shaped galaxy measuring 100,000 light-years (587,800,000,000,000,000 miles) from edge to edge. Thirty thousand light-years (176,340,000,000,000,000 miles) from the center of this galaxy, the planet Earth travels in endless revolution around an average star called the Sun. Joining Earth in her travel through space are eight other planets which make up the solar system. The entire system speeds through space, traveling toward Vega, a distant star in the northern heavens. Only 50 years ago, this might have been the beginning of a tale out of science fiction. But this is not fiction. It is real.

From the earliest times, people gazed in awe at the heavens. They marveled at the rains that washed the Earth, the clouds that floated across the sky and the lightning that seemed to split the heavens. But more than all of these, the brilliance of the stars truly captured the imagination of Mankind down through the ages. What were these lights in the night-time sky? What caused them to appear only in the darkness? What was the Sun that warmed the Earth and caused crops to grow? What was the moon that seemed to control the tides and seasons? It is no wonder that most of the early religions worshipped gods of the Sun or moon. No other phenomenon carried quite the mystery or the magic of these strange heavenly bodies.

Worship, however, was not enough to explain why the Sun rose every morning out of the east and set every night in the west. Nor did it explain the cycles of the moon or the movement of the constellations. Mankind has always been eager to know. In quest of knowledge about the heavens, the science of astronomy was born. Fifteen hundred years before the birth of Christ, people were observing the phases of the moon and the appearance of the planet Venus. It is true, of course, that these observations were used to predict various events that would take place on Earth. (Man has always believed that the stars somehow

Right: A 19th-century print of Chinese bird and dragon kites. The Chinese were interested in flight. Previous spread: The Apollo 4 Liftoff. The Apollo 4 space mission was launched at 7:00:01 am (EST), 9 November 1967. The objectives attained by the Apollo 4 Earth-orbital unmanned space mission included: (1) flight information on launch vehicle and spacecraft structural integrity and compatibility, flight loads, stage separation and subsystem operation, and (2) evaluation of the heat shield under conditions encountered on return from a moon mission.

control his destiny.) Regardless of the reason for the observations, Babylonian priests accumulated an amazing amount of information about the periods of the Sun, the moon and the planets. From all of this, they developed calendars and sun clocks. Oddly enough, these ancient observations are of much interest to astronomers today, for they act as historical check-points for astronomical data.

Babylonian astronomy influenced the ideas of civilizations that followed. The mathematician, geographer and astronomer, Ptolemy, knew and studied Babylonian charts and calendars. Ptolemy lived in the second century AD, a time when ideas about the universe were changing. He had his own ideas to put forth. His scheme of the universe seemed so flawless, in fact, that it became accepted cosmology for nearly 1400 years. There were others, many others—Greeks who proved the Earth was round, who even measured the distance around the Earth using geometry and the latitude between two cities. Others catalogued the stars. Some of these ancient astronomers went so far as to claim that the Earth was not the center of the solar system. This was an opinion in direct contrast to the thoughts of Ptolemy. He believed that the entire universe was a sphere turning on an axis with the Earth as its fixed center. The Earth, he thought, was a flat sphere. The sun, moon, and all the planets revolved around this sphere. Man, according to Ptolemy, was the center of the universe.

In all of these observations and studies, in the worship and the wonder, in some of the earliest art and legends, there was always the desire and the yearning to travel among the stars. But how could anyone escape the bonds of gravity and soar through the sky? No one knew and all but the most tenacious dreamers believed it beyond the realm of possibility. But the folk tales persisted. There was the tale of the Babylonian king who flew to heaven on the back of an eagle, the myth of Icarus and Daedalus who tried to fly with feathered wings, and the Chinese legend of the 'flying carts.' The desire and the dream were rooted deeply in every civilization of the world.

Then in 1232 AD, Mongol hordes swept down out of the North to attack China. They were repulsed by a strange weapon the Chinese called the 'fire arrow.' They were not, as we might think, flaming arrows. They were, instead, the primitive ancestors of modern rockets. Filled with an explosive combination of saltpeter and black powder, fire arrows could be 'launched' to hit more distant targets than an ordinary arrow.

The world took little notice of this event. The black powder, of course, became useful as gunpowder. Fire arrows became rockets, useful as signals, weapons and for sending out lifelines in sea rescues. They achieved their greatest popularity as fireworks. This was as far as the rocket evolved for hundreds of years.

In the twentieth century scientists took another look at the rocket. They began to make a careful study of just how rockets worked. For the first time, men began to understand the principles and the potential of the rocket. As a teenager in Russia, Konstantin Tsiolkovsky became interested in space travel. The idea stirred his imagination. He read everything he could find on mathematics, astronomy and physics. On his own, he developed a basic theory of rocket propulsion. In 1898, he wrote and submitted an article entitled, 'The Investigation of Outer Space by Means of Reaction Apparatus,' to the magazine, *Science Survey*. The article, however, was overlooked and was not published until 1903. Undiscouraged, Tsiolkovsky continued to write both technical papers and science fiction.

It was not until Hermann Oberth of Germany published an article in which he described the application of rocket propulsion to space flight, that Tsiolkovsky received recognition.

During the time that Tsiolkovsky and Oberth were writing and working on the idea of rocket propulsion, Robert Goddard, in the United States, was also studying the feasibility of rockets for high altitude flight and space exploration. Goddard, unlike Tsiolkovsky and Oberth, chose to test his theories through actual experiments. A great deal of his time was spent in devising the hardware to carry out those experiments. In 1914, he was awarded patents for a rocket which used solid and liquid propellants and a multistage rocket. It was Robert Goddard who designed, built and flew the first successful liquid fueled rocket. It was launched on 16 March 1926 at Auburn, Massachusetts. This primitive craft flew 184 feet (56 m) in $2\frac{1}{2}$ seconds. As he worked, Goddard developed many ideas which came into practical use with later large rockets—such now familiar items as liquid-fueled, self-cooled motors, gyroscopes for guidance and control, fuel pumps, reflector vanes which helped to stabilize and steer the rocket and parachutes which allowed the rocket to be recovered. Goddard took care to patent all of these features. It

was almost impossible not to infringe on one or another of his patents, as those who were to follow him discovered.

Goddard could have had a position of leadership within the infant space industry. But he had suffered some bad publicity early in his career when he submitted a paper titled, 'A Method of Reaching Extreme Altitudes,' to the Smithsonian Institution. He was deeply suspicious and suspected that others were plagiarizing his work. So he continued to work alone, almost secretly. He published very little of his experiments and accomplishments. Even when members of the American Interplanetary Society tried to contact him, he proved uncommunicative. The California Institute of Technology Rocket Research Project also tried to persuade Goddard to join them. He might have done so but one stipulation required that all ideas and projects must be mutually open. Goddard was uneasy with those terms and did not join the project.

Top: Robert H Goddard at Clark University mounting a rocket chamber for use in his experiments of 1915–1917.
Above: Goddard (second from left) and his four-chamber rocket—7 November 1936.
Left: Dr Robert H Goddard with a combustion chamber and rocket nozzle.

11

Dr Robert H Goddard's rocket (left) in flight over Roswell, New Mexico, 19 May 1937. It reached an altitude of some 3250 feet.

The launch of Dr Robert H Goddard's rocket flight of 26 August 1937 near Roswell, New Mexico. It had gimbal steering as well as catapult launching.

Robert Goddard continued to work alone, stymied by a lack of funds which the Rocket Research Project might have provided. The lack of publicity that Goddard's work received also held back US space research. Because so much of his work was unknown, many of his experiments were being repeated by others. Goddard was never to attain his original goal, that of using rockets for upper atmosphere research. But his theories advanced in the 1916 Smithsonian paper were to prove well-founded and today his work is honored by NASA's Goddard Space Flight Center.

In 1945, the Jet Propulsion Laboratory built and launched a rocket designed specifically for upper atmosphere research. The rocket, named the WAC-Corporal, was launched on 26 September 1945. In that first flight, it reached a height of 43 miles (70 km). The Jet Propulsion Laboratory was to continue its research with a larger, more sophisticated rocket called the Aerobee. It became one of the most useful rockets in the US high altitude research. New light was being shed on the pathway to the stars.

Curiously enough, neither Goddard's tireless work nor that of the Jet Propulsion Laboratory really gave the impetus to the US space program. Instead, a group of German scientists were responsible. When World War II came to a close, the US Army captured a large number of German V-2 rockets in an underground factory located in the Harz Mountains. Captured with the rockets were several key German scientists, among them Walter Dornberger and Werner von Braun. The Germans had discovered and reinvented for themselves much of what Goddard already knew. But they had gone beyond what he had accomplished. Financed by the German war effort, the V-2s were the most sophisticated missiles known.

The captured V-2s were brought to the US to be tested by the Army at White Sands Proving Grounds in New Mexico. Realizing the potential for scientific research, the Army allowed interested groups to use the rockets for high altitude research. Several university groups took advantage of the opportunity. Because of this foresight, the V-2s produced a broad range of non-military data.

With the end of World War II, a group of scientists and engineers from the Communications Security Section of the Naval Research Laboratory were looking for new research projects to tackle. At a meeting, one of the suggestions was that the group might convert its wartime knowledge of missiles and communications to a study of the upper atmosphere. Because this was the eighth to go on the blackboard, it became known as Project 8.

There was much enthusiasm for Project 8. When it became officially adopted as a new field of research, both engineers and scientists had something to cheer about.

Above: Eight of the 12 members of the National Advisory Committee for Aeronautics—Langley Field, Virginia, 1934. Charles A Lindbergh is seated at the far left and Orville Wright is seated fifth from the left.
Below: The Tiamat—the first experimental vehicle launched from Wallops Island 4 July 1945.

Physicists, intrigued with rocket flight, considered the field one of greatest importance. Engineers found the idea of instrumenting and launching the rockets challenging. The project had another point as well. Because knowledge of upper atmospheric conditions was vital to communications as well as the design and flight of missiles, it seemed likely that the Navy might support the project.

The proposal for the project was approved in December of 1945. At this point the fun began. No one in the group had any experience in upper atmospheric research. It was time for some intense schooling. Aerodynamics, telemetery, rocket propulsion—these were the names of some of the courses. Each member of the group read and studied and then lectured his fellow researchers.

They were now formally known as the Rocket Sonde Research Section. Their goal was to study the upper atmosphere. The problem was how to launch the instruments that were needed. While they were considering use of the Jet Propulsion Laboratory's WAC Corporal, word came of the V-2s at White Sands. The news was both literally and figuratively a boost to the infant Rocket Sonde Research Section.

The V-2 was particularly suited to the type of research the group wished to carry out. For one thing, it was available immediately. For another, it could reach an altitude of 100 miles (160 km) while carrying a payload of a metric ton. This exceeded the capacity of any other available rocket. It also meant that investigations of the ionosphere would be possible at the outset. The large weight-carrying capacity was a particular boon since it meant that instruments and equipment would not have to be pared down to fit a restricted payload. Of further value for upper atmospheric research was the fact that the missiles would have to be fired in an almost vertical trajectory due to a very narrow range.

There was much to do, however, before the first rocket soared from the White Sands Proving Ground. And it had to be done very quickly, for the rockets were scheduled to be launched in rapid succession. The existing German warheads were unsuitable for carrying scientific instruments. So the Naval Research Laboratory provided nose cones

Right: A Tiamat in flight over Wallops Island in August 1945. The Tiamat was part of a project for developmental testing of the US Air Force's first air-to-air missile. The vehicles tested at Wallops contained all the elements of an actual missile except a warhead and guidance system.

14

designed especially for housing research equipment. Telemetery equipment was also furnished and ground stations were erected at the White Sands range. There were hundreds of details to be worked out. Each missile must be equipped with radar which would provide information through telemetry. Each rocket must have a radio receiver to abort the flight, should a rocket begin to falter after launching. Cameras and other optical instruments must be set up on the range. Still other cameras, photocells and magnetometers to interpret the measurements of such things as aerodynamic pressure, cosmic ray fluxes, and the direction of the sun were needed. A whole series of new techniques had to be developed. To obtain some of the information, the spent rockets must be recovered. In some cases parachutes were useful. In others, explosives were used. This destroyed the streamlining tendency of the rocket and caused it to float to the ground. (The scientists referred to this as 'maple-leafing.') Even the sound ranging techniques used in World War I were brought into play.

The drive toward space exploration was suddenly burgeoning, growing like a mushroom. In these early years, scientists were learning just what a complex venture they had embarked upon. As with every new venture, there were details which had not been considered—tracking, timing signals, safety, telemetry, range communications, aborting flights, radio frequency interference, recovery of records and instruments. It seemed and was almost endless. There were other problems to be solved in the assembling, instrumenting, fueling and testing of the missiles—not to mention all the things that could go wrong at launch.

But here was a beginning. The door to space exploration was swinging open. And the view beyond looked promising. The accumulation of information was increasing steadily. The last V-2 was fired in 1952. By that time, tremendous amounts of information had been gained about atmospheric temperatures, pressures, composition, densities, ionization, winds, the magnetic field, atmospheric and solar radiation and cosmic rays. All of this information would prove invaluable in the coming space age.

Scientists were beginning now to look for a smaller rocket to carry on their research. The V-2 had been excellent as a tool when everyone was learning. But smaller rockets offered the advantages of being cheaper to produce and simpler to assemble and test.

Above: Hermann Oberth (foreground) with officials of the Army Ballistic Missile Agency at Huntsville, Alabama, in 1956. Left to right: Dr Ernst Stublinger (seated); Major General H N Toftoy, Commanding Officer; Dr Eberhard Rees, Deputy Director, Development Operations Division; and Dr Wernher von Braun, Director, Development Operations Division. Below: Dr Wernher von Braun, one of the key developers of the World War II German V-2 rockets.

Smaller and simpler rockets could be tested at places other than the White Sands Proving Grounds. So it was that the V-2 panel began to develop other types of single and multistage rockets.

Since the V-2 Panel represented scientists from the military, the universities and such companies as General Electric, there was almost unlimited talent for rocket development.

At the Applied Physics Laboratory were James Van Allen and his colleagues. With support from the Navy's Bureau of Ordnance, they undertook to develop the Aerobee rocket. Meanwhile, the Naval Research Laboratory was developing a large rocket called the Neptune. The name, however, had to be changed since a Neptune aircraft already existed. So it was rechristened the Viking. The Viking was actually a rocket ahead of its time. Since everyone was aware that the supply of V-2s could not last forever, the Viking was essentially developed as a replacement. For the time, it was the most efficiently designed rocket in existence. On a test flight in May 1954, the Viking reached a height of 159 miles (255 km) above the Earth's surface. But for all of its impressive range and efficiency, the Viking had one big drawback. It was very expensive. Since most of the groups working in rocket research had budgets of only a few hundred thousand dollars per year, the Viking was beyond consideration as a tool for research. There was still the Aerobee. Because of the success of both the Aerobee and the Viking, other types of rockets began to appear. Each was produced with definite goals in mind. First, of course, was the matter of keeping the cost down. Scientists all agreed that a simpler type of rocket would be a great advantage. Further, they wanted to attain greater altitudes and carry more payload. It would also be a definite advantage not to be restricted to the range at White Sands Proving Grounds. Great ingenuity and determination went into the development of these new rockets. They were launched from various geographical locations—even from balloons. (James Van Allen playfully called these rockets 'rockoons.') It was a rockoon which sent back information from the Arctic about a zone of soft radiation in the upper atmosphere. Later the belts of radiation in this zone were named the Van Allen Belts. In 1957, another balloon-borne rocket, the Far Side, established a high altitude record of 4000 miles (6400 km). (It had been carried by balloon to a height of 19 miles (30 km).

The panel, meanwhile, was increasing the scope of its activity. It was evolving from the V-2 Upper Atmosphere Research Panel to the Upper Atmosphere Rocket Research Panel or the UARRP. This name was fine until the advent of the International Geophysical Year. The International Geophysical Year was a program of world-wide research which was initiated in July 1957 and continued through December of 1958. At this time, the panel changed its name once more to the Rocket and Satellite Research Panel.

The actual panel had a very small membership, restricted to working members only. The number of representatives from the various agencies was also limited. The membership felt that this was the only way to keep the meetings from becoming unmanageable. They did not wish to discourage interest or participation in the meetings, however. So a band of interested observers was always allowed to attend and to join in the discussions. One of the groups which fell into this classification was the National Advisory Committee for Aeronautics.

The panel had no official status whatever. In actuality it was rather like a club. Its membership represented every institution engaged in rocketry research. But, adding up members and observers, there was a substantial number of people involved. Despite its loose organization and lack of official charter, the panel was the prime source of expertise in rocket research. Various government agencies relied on it for information. Various military branches also used the data gained by the panel. The National Advisory Committee for Aeronautics and the Defense Department's Research and Development Board both relied upon the studies of the panel.

With the advent of the International Geophysical Year or IGY as it was called, a new idea was advanced. Originally, the panel had planned only a sounding rocket program for the IGY. Now enthusiastic advocates were suggesting the launching of satellites. Members of the panel liked the idea. To explore the subject more fully, they sponsored a symposium at the University of Michigan in January 1956.

Interest in satellites grew like wild fire. At some point amid all the discussion, a suggestion was made that the United States should establish a separate and permanent space agency. After the launching of the

Russian satellite, Sputnik I, on 4 October 1957, the panel began to push the idea of a civilian 'National Space Establishment.' Clearly the time was right for such an agency. Americans were both awed and concerned over the successful launch of Sputnik. Space probes and atmospheric investigations were no longer merely a matter of scientific study. If a satellite could be launched to orbit the Earth . . . then wouldn't it be possible to launch a spaceship carrying a man? Shouldn't the US be equally as involved in a space program as Russia?

Members of the panel drew up a series of recommendations together with a plan to visit various congressmen and officials who could aid them. They called on the vice-president, the Academy of Sciences, the staffs of House and Senate committees, the general manager of the Atomic Energy Commission, the US Information Agency and others. Members of the Joint Committee on Atomic Energy were shocked by the cost of the proposed program—over a billion dollars a year!

All of this led to the National Aeronautics and Space Act of 1958. The National Advisory Committee for Aeronautics would be absorbed by the new National Aeronautics and Space Administration. No one realized at the time, least of all the panel, that NASA was to become a household acronym.

When the National Aeronautics and Space Administration came into being in 1958, it was fortunate to inherit over a decade of research by the Rocket and Satellite Research Panel and important contributions of the International Geophysical Year. Because of all the vital data which had been compiled, NASA was off to a flying start. Under the auspices of the International Geophysical Year, America had launched her first successful satellites, Explorer 1, launched on 31 January 1958, and Vangard 1, launched 17 March 1958. Amazingly, Vangard 1 still transmits via a solar battery. In addition to the Explorer and Vanguard series, there were the Discoverer group, Tiros, a weather satellite, Midas, whose letters stand for Missile Defense Alarm System, Samos 2, which was a miliary reconnaissance satellite and Echo, whose aluminum skin reflected the sunlight so brilliantly that it was visible as a star in the night sky. These and others preceeded the first manned satellite into space.

President John F Kennedy in his historic message to a joint session of the United States Congress on 25 May 1961 declared 'I believe this nation should commit itself to achieving the goal, before this decade is out, of landing a man on the moon and returning him safely to Earth.' This goal was achieved by Astronaut Neil Armstrong, the first man to set foot on the moon, at 10:56 pm, Eastern Daylight Time, 20 July 1969.

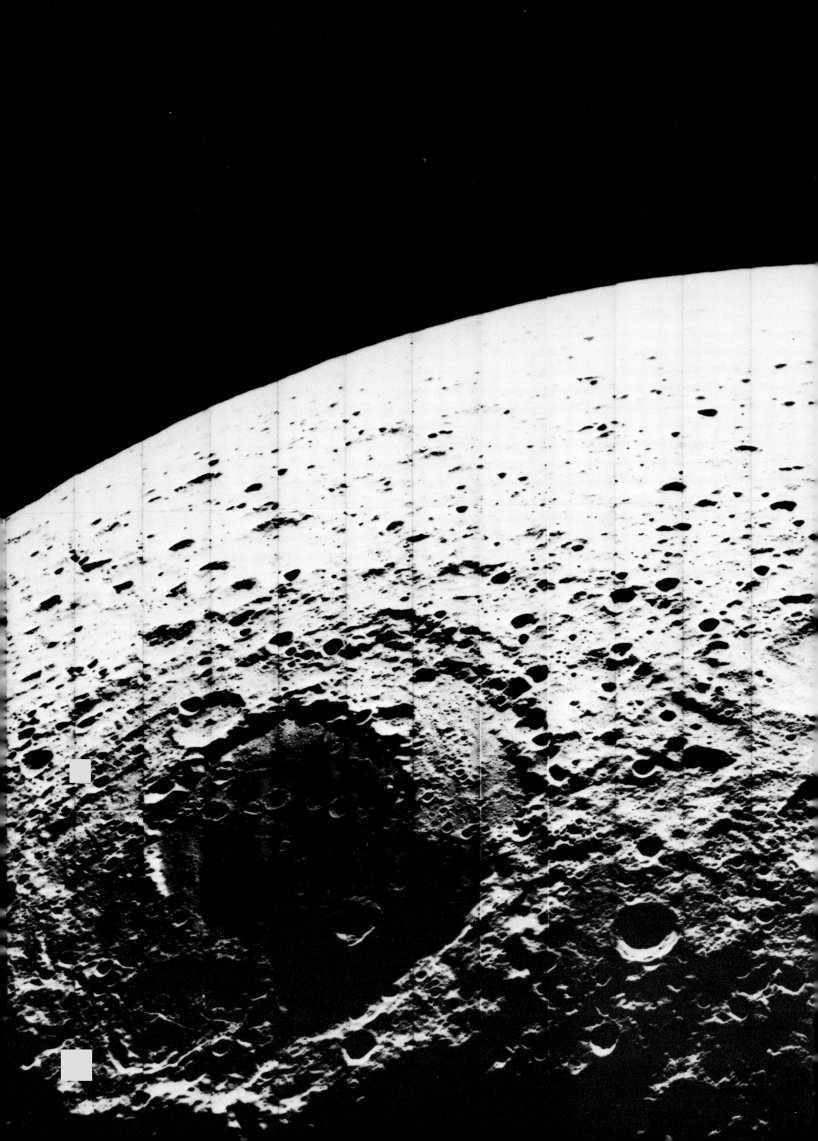

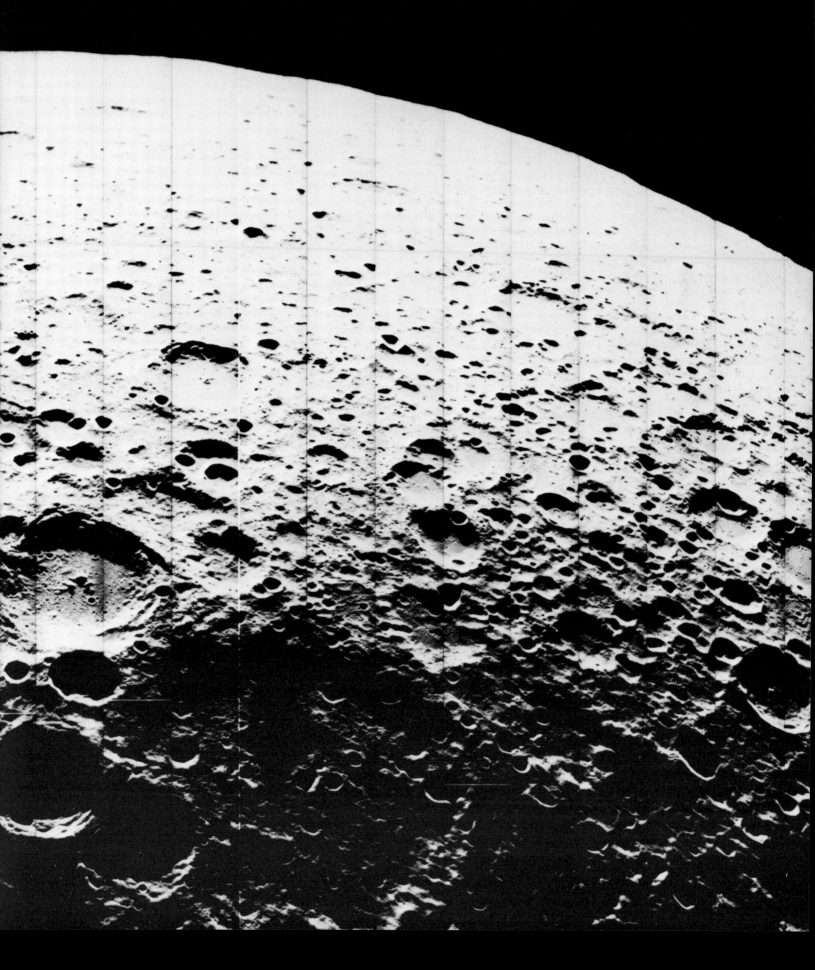

Aim at the Stars

In 1958, the year of NASA's formation, the US boasted an impressive array of sounding rockets. They ranged from the small Deacon, Cajun, Arcon and Arcas rockets to the medium-sized Aerobee and Aerobee-Hi rockets. Largest of all were the Vikings which had replaced the V-2s. The small rockets all used solid propellants, while the second-stage of the Aerobee and the Viking used liquid propellants. It was discovered that multistage combinations such as the Nike-Deacon and the Nike-Cajun could achieve higher altitudes with greater economy than the single-stage rockets. There had been a variety of launches—from land, sea and balloon. There had been almost inexhaustible testing. All of this gave NASA an immediate, on-going space program.

NASA was particularly fortunate to inherit such large rockets as the Jupiter C. It was used in combination with the Redstone missile to launch the first US satellite, Explorer 1. Many of these vehicles had been created for use by the military. The Redstones, for instance, were originally created for the Army by Werner von Braun and his team. The Vangard came from the Naval Research Laboratory. And there were the Thor and the Atlas created by the Air Force. The presence of all this existing technology gave NASA more than a head start.

Thus it was, a bare three years later, that NASA was preparing for the first manned space flight. The first 'astronaut' for the US, however, was not a man but a charismatic little chimpanzee known as Ham. A group of these animals had undergone training at Holloman Aeromedical Laboratory in New Mexico. Trained as rigorously as their human counterparts, they learned to experience weightlessness and acceleration, to pull levers in response to flashing lights and to deal with the isolation that space flight required. Ham handled his first space flight with aplomb, proving that manned space flight was certainly possible.

With Ham's flight, the future of Project Mercury was assured. Now, manned testing was begun. On 5 May 1961, a Redstone rocket lifted a Mercury Capsule to a distance of 300 miles (480 km) and a height of 113 miles (180 km) above the surface of the Earth. The capsule carried astronaut Alan B Shepard Jr.

It was a moment of the highest drama, that first flight. It was a moment when suddenly Man was face to face with the enormity of what was about to take place,

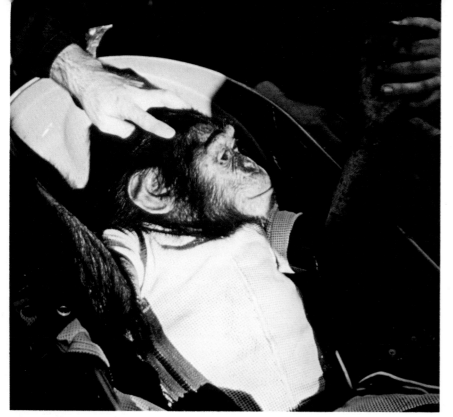

Above: Ham, the first chimpanzee ever to ride a rocket, is given a post-flight physical check after a 5000-mile ride in a Mercury-Redstone 2.
Opposite: The Mercury launch of 29 November 1961.
Below: NASA Astronaut Alan B Shepard Jr unsuits for a physical examination after his sub-orbital flight 6 May 1961.
Page 18–19: A photograph of the moon by the Lunar Orbiter, 17 October 1967.

a moment when centuries of impossible dreams were about to become reality. Behind all the drama were months and years of tests and exploration, miracles of engineering and research, and now, the slow and tedious countdown. 'T minus 25 seconds.' The umbilical tower dropped away from the Redstone rocket. Then the announcer's voice came, tight with excitement, through the PA system: 'Ignition . . . Mainstage . . . Liftoff!'

From the capsule came a calmer voice with the reply, 'Roger, liftoff and the clock is started.'

The clock would give the astronaut a precise reading for the start and termination of each function that he would perform.

Crowds of eyewitnesses strained their eyes against the morning sky to track the progress of the flight. Many found their eyes filled with tears and their lips moving in prayer for the gallant young man in the capsule. In an effort to ease what was no doubt an emotion-packed moment, the voice of astronaut Deke Slayton came over the capsule communicator's console. 'OK Jose, you're on your way!' The reference was to a TV comic called Jose Jimenez who sometimes played an astronaut.

The capsule continued to rise, slowly at first and then faster and faster. Mach 1, Mach 2, and still faster, Mach 3, Mach 4, the capsule and its booster were now beyond the maximum aerodynamic pressures. Now they passed through layers of the atmosphere where no man had ever traveled before. The gravitational pull was growing within the capsule, 2 g's, 3 g's, 4 g's, pressing Alan Shepard close into his molded couch. Five g's, and then at T plus 142 seconds, the booster cut away. The roar of the Redstone silenced, the explosive bolts fired and the clamp ring fell away. The posigrade rockets, tiny by comparison to the giant Redstone rocket, were activated. With their gentle 20 feet per second (6 m per second) push, the g-forces were gone. The capsule and the booster had separated. The flight controllers at Mercury Control Center drew a deep breath. The most dangerous moment in the flight was over. Traveling at a rate of 6400 feet per second or 4375 miles per hour (1950 m per second or 7000 km per hour), Shepard and the Freedom 7, as his space capsule was named, were proving the possibility of space flight.

At T plus 157 seconds, the capsule was beginning its turnaround. The men in the Control Center droned their instructions,

silently savoring the triumph. The automatic system devised for stabilizing the capsule was working out exactly as planned. The capsule now traveled big end forward, with the heat shield and the retrorockets pointing along the flight path. Shepard was now experiencing the condition of weightlessness. There was no discomfort whatsoever.

Then came a test of the manual controls. 'Switching to manual pitch,' came the pilot's voice. 'Pitch is OK. Switching to manual yaw.' A second and greater triumph; the manual controls worked as well as they had in any of the tests.

T plus 240 seconds and now Shepard checked the view through his periscope. One hundred miles (160 km) below rotated the planet Earth. Land masses, cloud cover, could he identify them? 'What a beautiful view!' he called back to those Earth-bound below. Could he see the moon? No, it had set. The launch was late. There was actually little time for sight-seeing, for the controls demanded his attention.

T plus 300 seconds: time for the retrorockets to fire. The retrorockets slow the capsule. They are not really needed on this first flight, but NASA wanted to test them, anyway. 'Retro attitude on green,' came the pilot's voice. 'Retro one, very smooth . . .

retro two . . . retro three. All retros are fired.'

T plus 360: the burned-out retrorockets must be jettisoned. But where is the retro-jettison light? There is no light. The straps are falling away. There's a noise but no light! The pilot tries the override button. And there is the light. Retro-jettison has occurred. A sigh of relief escapes Shepard's lips. This was the only malfunction he was to encounter during the whole flight. Actually it was also something of a triumph for those engineers who had planned for just such an emergency.

T plus 470 seconds: The .05-g light comes, signifying reentry into the Earth's gravitation. The g's are increasing; to prevent loss of consciousness, Alan Shepard tenses his muscles, straining at the harness that confines him to his couch. The forces are stronger on reentry and Shepard's voice reflects the strain. They reach a maximum of 11 g's. Then, with the pressure eased, he called, 'OK.' This was the most severe physical stress endured in the flight and Shepard handled it easily. It was actually much less than he'd endured in the centrifuge back at Langley Air Force Base.

The capsule is beginning to glow with the heat of reentry and the temperature in the cabin reaches 111 degrees F (44°C). It was T plus 580 seconds: the telemetry signals

Above: President Kennedy congratulates Astronaut Alan B Shepard Jr on his historic ride in the Freedom 7 spacecraft. The ceremony took place on the White House Lawn. Below: Astronaut M Scott Carpenter suiting up.

were beginning to weaken. The capsule was descending rapidly. Voices from Mercury Control call out the altitude, fifteen, fourteen, twelve, eleven, main chute, ten thousand feet.

The main chute unreefs and it looks good, with the rate of descent reading 35 feet per second (11 m per second). Mercury Control has now lost contact with Alan Shepard because the signals are blocked by the curvature of the Earth. All is silent as the men listen to the recovery forces 302 miles (486 km) away from the launch pad. Now there is hushed rejoicing as they learn that the orange and white parachute has been spotted by the USS *Lake Champlain*. Then at T plus 922 seconds comes 'Impact. The capsule is in the water—looks good.'

Then came the words from Shepard himself, 'Everything is A-OK.' Those words were to become famous in headlines around the world. Everything was A-OK. America's first man in space had come through with flying colors. People around the world rejoiced and pinched themselves to make sure it wasn't really some far-fetched dream.

The doctors, plus astronauts Donald 'Deke' Slayton and Virgil 'Gus' Grissom hurried to meet Shepard at Grand Bahama Island. They were only a few minutes ahead of the plane from the Carrier *Lake Champlain*. All during the trip to the island, they looked at each other, grinning and shaking their heads in disbelief.

Alan Shepard had survived his flight in fine style. There was an intensive two-day medical study in the Bahamas. Shepard, however, was fine. In the words of one of his doctors: 'He didn't even need a bandaid.'

Then came the slow and inevitable letdown. The flight had been text-book perfect. None of the things the men at Mercury Control had worried about had happened. As relief replaced the months of tension, 'Gus' Grissom joked, 'There won't be this much fuss the next time. Once you've seen one, you've seen 'em all.' (Grissom himself was to man the next capsule into space the following 21 July 1961.) The flight was equally uneventful. But it added immeasurably to NASA's confidence in the space program.

Named for the first orbital flight were John Glenn and his back-up, Scott Carpenter. But before Glenn's historic orbit, a chimpanzee would test the capsule and the tracking stations. The chimp would make two orbits of the earth before splash-down in the South Atlantic.

Sixteen orbital tracking stations existed around the Earth. During the flight, each maintained contact with the others by means of open telephones. Each kept a record of flight data as the capsule passed within range of the station. The electronic equipment required for these stations was nothing short of mind-boggling. There were clocks to synchronize timing to one five-thousandth of a second, computers that could calculate and plot the orbit of the capsule halfway around the world in seconds, and the most accurate radar in the world.

Each range was set up through painstaking diplomatic arrangements with each country involved. It was important to make clear that the tracking range served purely scientific purposes. Five foreign nations were involved in the tracking network. They were Australia, Mexico, Bermuda, Nigeria and Zanzibar. Within two years the stations were set up and ready to operate. The orbit, which swung from roughly 32 degrees north to 32 degrees south, was designed to keep the capsule traveling as much over water or US territory as possible. Each orbit, however, would vary slightly because of the Earth's rotation.

Three orbits were planned for this first flight because further trips would carry the astronaut outside the Atlantic splash-down area. The Mercury capsule of that time lacked the cooling system or adequate electrical power for long distance flights.

This was another small part of the preparation for John Glenn's historic space flight around the world.

On 20 February 1962 the Friendship 7 lifted off from Cape Canaveral, Florida, into a flawless blue sky. Engineers in the Mercury Control Center could listen in as John Glenn reported to tracking stations around the Earth. The flight was going well, except for a small problem with the automatic controls. Glenn was compensating for this by flying the ship manually.

The technicians in the back room told a different story. Radio signals from the spaceship showed that the landing bag, which would act as a cushion when the capsule hit the water, had opened prematurely. If the signals were true, this also meant that the heat shield, attached to the landing bag, had also come loose. The ship would not be protected against the blistering heat of reentry.

While Glenn's ship passed from daylight to darkness and back to daylight again,

A Project Mercury spacecraft carrying Astronaut Virgil I 'Gus' Grissom on NASA's second manned space flight—21 July 1961. He landed in the Atlantic Ocean about 305 statute miles from Cape Canaveral. The craft reached an altitude of about 118 statute miles and a speed of approximately 5310 miles per hour. The suborbital flight of the Mercury-Redstone 4 required about 16 minutes.

NASA's scientists tried to determine what action to take. Tense and anxious, they all agreed on one thing, 'Take no chances.' In the end they advised Glenn not to jettison the retrorockets. It was just a chance, but the retrorockets might hold the heat shield in place. To those tracking the flight at Mercury Control Center, the waiting grew almost unbearable. Radio communications had blacked out as it was known that they would during the intense heat of reentry. But what about Glenn? What was happening up there?

The complete story of John Glenn's flight would not be known until hours later. Seconds before liftoff, John Glenn lay strapped in his capsule; 70 feet (27 m) above launch pad 14, he heard the final message from his fellow astronaut, Scott Carpenter. 'May the good Lord ride all the way.' Then followed the voice of Alan Shepard in the final countdown, 3, 2, 1, zero!

The rocket engines fire. The giant bird quivers, readying for her flight. And now, with a smooth upward surge, the craft takes flight. 'Roger,' Glenn reports, 'the clock is running. We're under way.'

The reply comes back in Shepard's familiar tones, 'Hear you loud and clear.' At the bottom of his window, Glenn watches the horizon turning. The Atlas rocket bears to the right heading. Vibration is building indicating the craft's encounter with maximum aerodynamic pressure. 'Little bumpy along about here,' he reports. There is a dull roar in his ears from the engines. Time now to start regular reports to the ground crew. Calmly he reports on the state of the fuel, oxygen, electrical system. From the ground, equally calm, comes Alan Shepard's voice, 'Loud and clear. Flight path looks good.'

The vibration has grown more intense. The whole rocket is shuddering. This is the maximum air resistance. Then suddenly the vibration drops. There is a collective sigh back at Mercury Control. The ship has made it through the period of greatest stress. The needle that indicates cabin pressure is dropping. Past seven pounds per square inch, it slowly inches downwards. Then it stops. 'Cabin pressure is holding at 6.1,' he reports.

His chest is suddenly growing heavy, weighing him down into his seat. Acceleration is growing and so are the g-forces. Before he breaks free of Earth's gravity, his weight will increase to nearly half a ton. He has experienced g-forces ranging as high as 16 at the training center so this does not especially concern him. The contoured

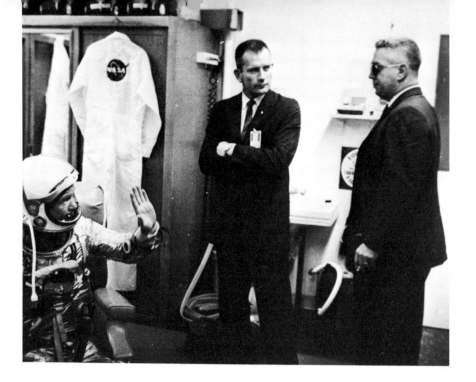

Above: Astronaut John Glenn talks with Mercury launch director Walter Williams (right) and fellow-astronaut Donald 'Deke' Slayton before his orbital flight—20 February 1962.

Left: Astronaut John Glenn's neck wrap is adjusted before his historic flight—20 February 1962.

Opposite: Astronaut John Glenn adjusts his space suit as he prepares to leave for the launch pad.

Below: The insertion of Astronaut John Glenn into the Friendship 7.

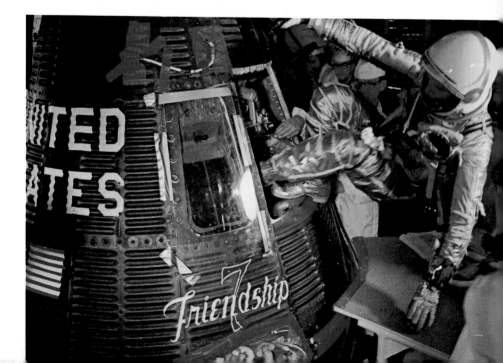

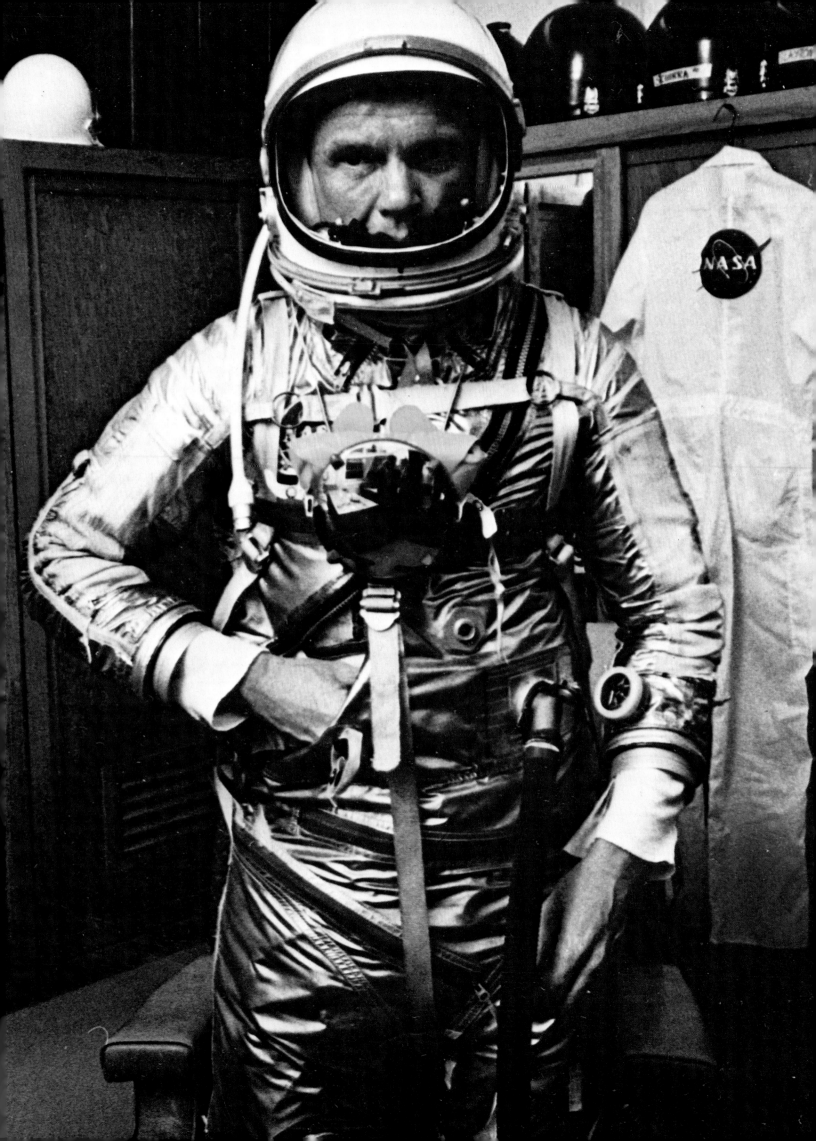

Above: Glenn as he climbed into the Mercury capsule atop the Atlas rocket at the Cape Canaveral Missile Test Center—20 February 1962.

Below: Glenn displays his three reflection mirrors, used to reflect readings on instruments in the capsule during space flight.

couch cradles his body in comfort. Two of the three main Atlas engines are jettisoned and the acceleration slows rapidly. The ship trembles for a moment as the rockets break away. Glenn, spotting the smoke through his window, thinks the escape tower has fired. 'The tower fired,' he reports back to base. He notices then that the smoke is from his engines which floated upward when the boosters broke away. The tower is clearly visible. Had trouble developed at this point, the escape tower would have carried the capsule out of danger. Once outside the atmosphere, it is no longer useful and will be jettisoned by the automatic programmer. Sure enough, like a dart with a long gray tail, the tower slips away.

Now the spacecraft pitches downward. Below is the Earth, the sky black above and white clouds scudding across the Atlantic. It is only a brief view before the craft pitches up again.

The powered portion of the flight is nearing the end. The fuel tanks are nearly empty. The Atlas rocket, whose stainless steel wall is thinner than a dime, becomes flexible. At this moment it is nothing more than a giant, stainless steel balloon. Bobbing with the motion of the flight, the empty rocket gives Glenn the sensation of being on a diving board. 'Seven, Cape is go; we're standing by for you.' It is Shepard's voice telling Glenn that everything on the ground is all right. 'Roger,' Glenn reports back, 'Cape is go and I am go. Capsule is in good shape. Fuel, oxygen, cabin pressure holding steady. All systems are go.' Its time to jettison the over-sized balloon.

The g-forces press down again. Then suddenly they decrease. Glenn has the sensation of pitching forward, somersaulting head over heels. It's a familiar experience from the training centrifuge. There is a sharp explosion as the Atlas is severed from the capsule and a burst of speed as the posigrade rockets take over.

He feels a slight lift, a sensation of floating. 'Zero g, and I feel fine,' he reports. The capsule is turning to assume the blunt-end-forward position normal for flight. Peering out the window, he can see the empty Atlas rocket just behind and a little above him. It seems to hang there in space, gradually dropping lower and moving farther away. He watches it, noting how long he can see it and from what distance. The information will be important to future flights when astronauts will be docking in space.

'Seven, you have a go, at least seven orbits.' Shepard's voice again carrying the welcome news. The Atlas has done its duty, Glenn's spacecraft is now in orbit 100 miles (160 km) above the Earth's surface. Its course could be maintained for seven orbits. Friendship Seven and John Glenn are now in control. Contact with the Cape is fading. The next transmission will come from Bermuda.

The Capsule Communicator on Bermuda is Gus Grissom. It's time for the orbit checklist. Switch positions must be checked at this point. Grissom also sends the schedule for firing the retrorockets. Timing is of the utmost importance to insure that the capsule will splash down near the recovery ship. As with every function of the space program, there are contingency recovery areas should anything go amiss. Grissom's voice is fading as the ship travels out of range. Ahead are the Canary Islands and the West Coast of Africa.

Out of the silence comes the voice from the Canary Island Station. 'Friendship Seven, this is Canary CAPCOM. Could you get started with your station report?'

Glenn has been retrieving a camera from his ditty bag, preparing to take pictures of the Earth's daylight side. 'This is Friendship Seven,' he responds. 'Stand by. I am getting out some equipment.' This is the first of a series of regular half-hourly reports. At this time, Glenn checks all switch positions and the readings on each of the major instruments. Then he must give the capsule's attitude, which is all-important for landing procedure. He completes the status check and glances out the window. The view is breathtaking. He reports, 'The horizon is a brilliant, brilliant blue. There, I have the mainland in sight at the present time. I have the Canaries in sight out through the window.' He had left Cape Canaveral only 18 minutes ago.

He has passed over the Desert of Africa and is moving toward Zanzibar on the East Coast.

Eastern Africa is covered by high, wispy clouds. He remembers that the US Weather Bureau has asked him to see whether he can determine the types of clouds and their heights from his altitude. He finds it is not difficult to discern the different cloud formations and their height can be easily reckoned from the shadows they cast.

He is now passing over Zanzibar. In addition to his regular report, there is his blood pressure measurement—120 over 80—just about normal. Its now time for some exercise. His doctors would like to know how exercise under weightless conditions compares with the same exercise back on the ground. Between Glenn's knees is a handle attached to an elastic cord. Grasping it with both hands, he pulls it to his chin. Thus extended it exerts a force of 62 pounds (28 kg). He releases the handle, letting it spring back to his knees. This exercise is repeated for 30 seconds at the rate of once a second. Now another blood pressure measurement. Blood pressure is now 129 over 74. This is an excellent reading and the doctors are pleased. There are yet other exercises to be performed. The Russian cosmonaut, Gherman Titov, had noticed dizziness and nausea in the weightless state. This occurred after six orbits in space and was particularly noticeable whenever Titov moved his head or body or watched rapidly moving objects. Glenn tries various head movements. He reaches about the cabin with either hand trying to touch switches and knobs. He tries throwing a beam of light rapidly, back and forth across the cabin. He notices nothing at this point.

The spacecraft is hurtling rapidly toward the Earth's shadow. People down there in Australia would call it sundown. In orbit, the perspective is different. In preparation, Glenn covers the cabin lights with red filters. The extra lights which are used for the camera are turned off. His eyes must become accustomed to the darkness so that he can see as soon as he enters the Earth's shadow. There's an eyepatch that he's supposed to put on to enable him to watch the colors of the sunset, yet at the same time accustom the other eye to darkness. But the patch won't stick. Glenn tries to adjust for this by keeping one eye closed. Glenn is now over the Indian Ocean, turning left to watch the setting Sun.

Using a miniature photometer with a polarizing filter, he is able to look directly at the Sun without harm. The Sun, approaching the horizon, is perfectly round. The horizon itself is a bright arc stretching as far as Glenn can see. Its brightness is made even more dramatic by the darkness of space lying above it, and the shadowy earth below. As the Sun disappears into the horizon, it seems to melt into a bright white band of light. Traveling at this speed, the Sun sets 18 times faster than on Earth. Wrapped in an orbital twilight, the horizon presents a striped ribbon of light extending from north to south. The bottom band is no longer white but orange. Gradually it fades

A view of Astronaut John Glenn inside the capsule during his flight. It was taken by an automatic camera on 16mm film. Glenn is scanning the instruments and viewing the Earth below him.

to red. Layers above this are darker colors. Finally there are layers of blue and then just the blackness of space. As the red fades, there appears a brilliant blue band at the horizon. This is what the astronomers who briefed Glenn before the flight will want to know. They feel it should be possible to study the Sun's corona far more effectively from space. Without the air to scatter the Sun's light and the possibility of blocking out the direct rays, the view of the corona should be spectacular.

As he checks his periscope, Glenn sees a bright light. It is the moon coming up behind his capsule. The light is reflected off the clouds below. The last blue band has faded from the horizon. Silhouetted clearly against the stars is the line of the Earth's horizon. One of Glenn's navigation devices is a chart showing his pathway through the constellations. A transparent slide fits over the chart. Along one side of this chart is a time scale. If he lines up the scale with the slide, he will be able to locate the stars which should be visible through his window. Future astronauts will use such charts to navigate their way by the stars, just as earthbound explorers have done for centuries.

What can that be? he wonders. He can see the Earth's horizon clearly in the darkness. But, perhaps seven or eight degrees above it is a glowing band shading from tan to buff-white. Stars, moving toward the horizon, seem to grow dimmer as they pass through the band. But once below it, they brighten before disappearing over the horizon. The band is well above the clouds, separated from them by a black area of space. This is a puzzle the astronomers will have to explain.

Below is Muchea, a station in Western Australia. The Capsule Communicator is Astronaut Gordon Cooper. 'Hello Muchea, we're doing real fine up here. Everything is going very well,' Glenn reports.

Cooper replies, 'You had an excellent cutoff, John. Your velocity was eight feet per second (2.4 m per second) low.' At a speed of 25,700 feet per second (7833 m per second), that averages out as an error of only 20 inches in a mile (32 cm per km).

Cooper calls again. 'Shortly you may observe some lights down there. You want to take a check on that to your right?' The lights are being turned on by the people of Perth and all the surrounding towns to serve as a beacon to the man in space. Glenn gazes out. Below are the lights of Perth.

From his altitude, they seem like the lights of a small town as seen from a high-flying aircraft.

'The lights show up very well,' Glenn reports. 'Thank everybody for turning them on, will you?'

The capsule travels on, across Eastern Australia, cloud-covered and nearly invisible to the astronaut. Australia is left behind and now the Pacific is below. At this point Glenn will attempt a transmission on his high frequency radio. The transmission can be received by the tracking station only if it is within 900 miles (1440 km) of Glenn's ship and on his side of the horizon. Once in a while, however, high frequency transmissions can be picked up half-way round the world. This is because they are very erratic. 'This is Friendship Seven, broadcasting in the blind to Mercury network, 1, 2, 3, 4, 5.' Fifteen seconds pass.

'This is Canaveral CAPCOM testing the HF. I did not read the capsule. Cape out.' Well, he thinks, they may not be able to get me, but I heard them loud and clear. Another 15 seconds pass. He should be hearing from Bermuda, the Canaries and the Atlantic Ship but all's silent. The silence is broken by transmissions from Kano, Nigeria and Zanzibar. Somebody down there is listening!

It is now time to eat. Until this moment, he hadn't thought of food and he realizes he's hungry. The food is contained in something like a large toothpaste tube. He opens it and squeezes. It tastes pretty good, apple sauce, he thinks. Eating is no problem, even in his weightless condition. He releases the empty tube and watches it float.

It's morning. He's crossed the International Dateline and time at this moment is going backwards. Yesterday was Wednesday. Today is Tuesday. The sun is coming up behind the capsule, a 'brilliant, brilliant red.' He hastens to put a dark filter over the periscope. The light of the sun through the scope lens is blinding. Glancing from the periscope to the window, Glenn is amazed to see hundreds of twinkling stars. His instruments indicate all is normal. But looking back, he discovers that the stars are actually bright specks that resemble fireflies!

Right: Astronaut John H Glenn Jr's spacecraft Friendship 7 being brought alongside the recovery ship after the splashdown following his world orbital flight—20 February 1962.

They glow with a mysterious yellow-green color. Close to the ship, they seem to be white. Glenn reports these mysterious particles to the station at Canton Island. They want to know if he can hear them hitting his capsule. 'Negative,' he replies. The particles swirl around the capsule, at times reminding Glenn of snow. As the sun rises, the particles are harder to see. Where once there were thousands, now there are only a few. Here is another space mystery.

The capsule is now over the coast of Mexico. But the cloud cover prevents him from seeing land. The ship seems to be drifting toward the right. Autopilot should hold it to the proper altitude, but it's still drifting. It has yawed to about 20 degrees. Then it is suddenly brought back on course. There is a throb of steam from the large attitude control thruster. This could be a problem. The big thruster uses so much of the hydrogen peroxide fuel that Mercury had not wanted to use it as the main attitude control. But it seems that the small thruster is stuck. If he has to use the large one exclusively, he may run out of fuel.

He's now over the Cape and Alan Shepard asks for a report on the problem. Glenn explains what is happening and says he will change to 'fly-by-wire.' In using fly-by-wire, Glenn will have to control the ship's attitude by himself. He would have to use the same electric signals as the autopilot used. This system will correct the roll and pitch of a ship. But, in order to correct the yaw, he will need to rely on the manual control for most of the remaining flight. It's a problem, but a small one.

The first orbit is complete. The craft swings into the second. He's once more over the Canary Islands. The autopilot still continues to give trouble. But now instead of drifting to the right, the capsule drifts left. To correct this, he makes a 180 degree turn. And there he is with the Earth coming toward him. The view is beautiful, much more pleasant than flying backwards as he has been doing. But much as he would like to, he cannot maintain the ship in its present attitude. So he resumes flying backward, peeking through the window at where he has been. Here is another sunset. The light from it reminds him of one of the arc lights out on the launch pad.

Far below to the north is a heavy electrical storm. The tops of the clouds glow ominously. Lightning flashes back and forth. The Weather Bureau Scientists will want to hear about this. They would like to improve the kind of information that the Tiros satellites send back to Earth.

Glenn is over the Indian Ocean. Here the tracking station is on board a ship. The ship relays a message from Mercury Control—

Opposite left: John Glenn's Mercury Friendship 7 space capsule with Glenn inside is pulled out of the water by men of the destroyer *Noa*, 20 February 1962. Glenn remained inside the capsule until it was on the deck of the ship.

Opposite right: Astronaut John Glenn is lifted from the destroyer *Noa* into a helicopter which took him to the USS *Randolph* at the completion of his three-orbital flight around the Earth.

Below: John Glenn checks his space capsule aboard the destroyer *Noa*. A technician inside the capsule pokes his hands through the escape hatch through which Glenn emerged from the space vehicle.

'Keep your landing bag switch in the off position.' The switch, Glenn notes, is in the off position. The problem he is having is with the attitude indicators. The instruments read normal, but a check out the window shows the craft has rolled 20 degrees to the right.

Another call from the station. 'Friendship Seven, will you confirm the landing bag switch is in the off position?' The call is from Cooper in Australia.

'Affirmative,' Glenn replies. 'Landing bag switch is in the center off position.'

'You haven't heard any banging noises or anything of this type?' Cooper questions.

'Negative.' This is what the Mercury Control hoped to hear.

The capsule is traveling through another sunrise. Orbital twilight spreads around the edge of the Earth. It brightens, slowly becoming a copy of the sunset band.

Here again are the firefly-like specks he encountered before. Twinkling like thousands of little stars, they drift slowly by the ship.

Another station checks in. 'Friendship Seven, this is Canton Island. We also have no indication your landing bag might be deployed.'

'Roger,' Glenn answers. 'Did someone report the landing bag could be down?'

'Negative. We had a request to monitor this and to ask you if you heard any flapping when you had high capsule rates.'

'Negative, I think they probably thought these particles I saw might have come from that.'

Beyond Canton Island, the next station on Kauai, Hawaii, checks in to ask if Glenn intends to go for a third orbit. The answer is affirmative. Eight minutes to cross the United States! Then he's talking with Bermuda. 'I can see the whole State of Florida just laid out like a map. It's beautiful. I can see clear back along the Gulf Coast.' At this point, his flight plan requires him to check the weather in the reentry area. Skies over Bermuda look beautiful—no problem for recovery, he reports to Grissom.

Glenn is traveling through his third orbital sunset. Somewhere below is Johannesburg, South Africa. But heavy cloud cover prevents him from seeing anything. Off on the horizon he can see lightning flashing. It's interesting to see a storm from the topside.

'Friendship Seven, this is the Indian CAPCOM. What is your control mode?' This is the Indian Ocean Station which is now contacting Glenn. He reports back that he is on automatic pilot, but that it 'is operating very erratically.' He is continuing to back it up with manual controls in an effort to maintain the attitude for retrofire.

Another sunrise, its ribbons of color lining the horizon, glimmers over the space craft. Glenn looks to see if he can spot the strange space particles of the past two orbits. They are there, but in the bright sunlight in front of him, he cannot see as many. They seem to be coming from ahead of the ship. He can see them spread out all over.

A voice on the radio interrupts his observations. 'Friendship Seven, this is CAPCOM. We are not receiving your transmissions.' It's the Canton Station. Glenn explains that he's been busy getting ready for retrofire. Lights must be set up and all the equipment must be stowed. The bright lights reveal that his instrument dials are still inaccurate. Glenn adjusts his attitude by the view through the window.

Another message is coming over the radio. 'Friendship Seven, Hawaii CAPCOM. We have been reading an indication on the ground of landing bag deploy. We suspect this is erroneous. However, Cape would like you to check this by putting the landing bag switch in auto position and see if you get a light. Do you concur with this?'

There is a pause while John Glenn considers this. 'Okay, if that's what they recommend, we'll go ahead and try it. Are you ready for it now?'

'Yes, when you're ready.'

'Negative, in automatic position did not get a light, and I'm back in off position now.'

'Roger, that's fine. In that case re-entry sequence will be normal.' Time is becoming critical. The capsule is nearing the California tracking station. Eight minutes to cross the United States and then splash-down!

'Friendship Seven, this is California, reading you loud and clear.'

'Roger, this is Friendship Seven. My capsule elapsed time is 04 plus 31 plus 35 on my mark 2, 3, 4, mark. Will you relay immediately to the Cape? I think we're several seconds off. Over.'

'Roger, we have you on that. Will give you the countdown for retro-sequence time, John. You're looking good.' The tension for the astronaut is building. Ground control is also experiencing tension but the calm encouraging radio transmissions mask the anxiety.

Glenn notes 50 seconds to retrograde. The California station gives him a mark

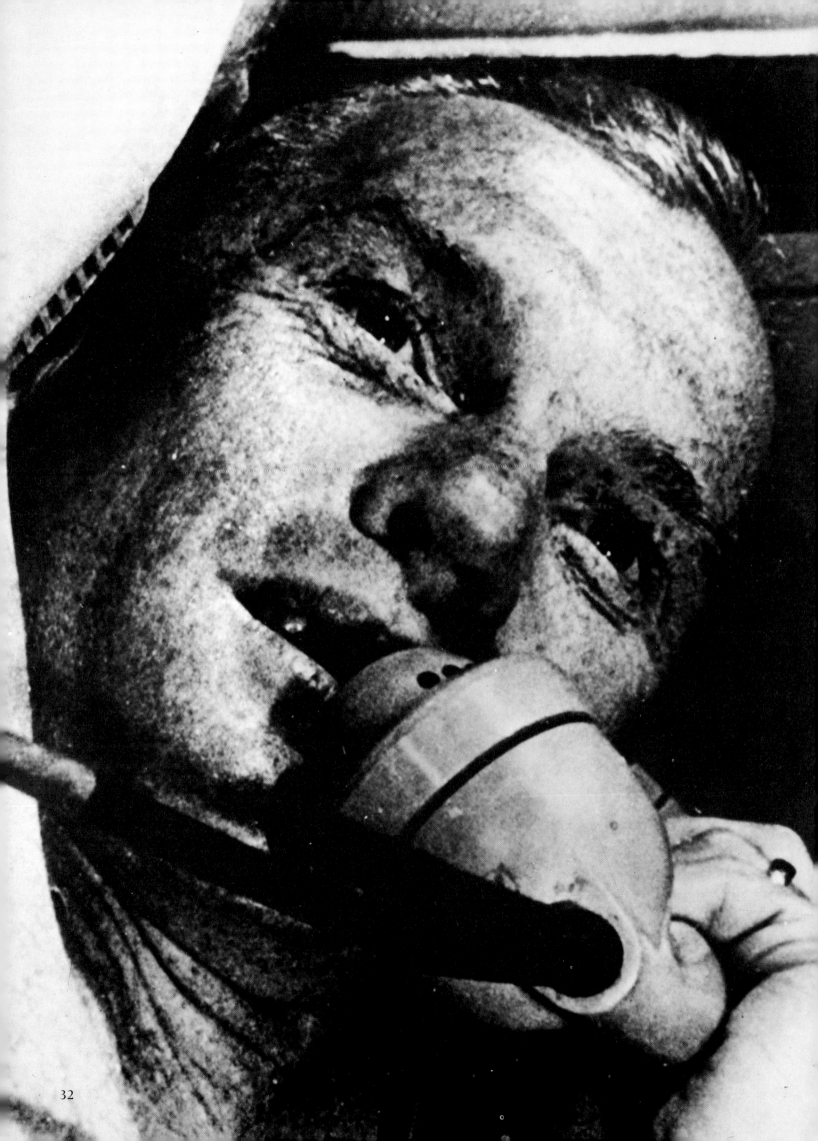

and he checks his clock again. Right on time, he sees. There is a buzz and a flash of yellow light to his right. A warning light, this tells him that retrofire will begin in 30 seconds. Fifteen seconds; now another light on the instrument panel flashes green. This is the retrosequence light. Below, CAPCOM is counting, 'five, four, three, two, one, fire!'

A sudden roar indicates the retrorockets have fired. The jolt takes Glenn by surprise. The force makes him feel as though he's heading back toward Hawaii. The second fires five seconds later, and then the third. These rockets will slow his flight but only slightly. He is still traveling at more than 17,000 miles per hour (27,200 km per hour)! The ship is holding its own, though the autopilot is still not working properly.

Texas CAPCOM relays a message from the Cape. Leave the retropackage on through the entire re-entry. Glenn will have to override the .05 g switch. The Cape is still not sure that the landing bag has not deployed. It is for this reason they request that the retropackage not be jettisoned. If the landing bag had been deployed, the heat shield would also have been freed. During re-entry, the capsule would be subjected to a force of 8 g's. The friction of the air would cause the capsule to incandesce with a temperature almost five times that of the surface of the Sun.

'This is Friendship Seven. Going to fly-by-wire. I'm down to about 15 percent [fuel] on manual.'

'Roger, you're going to use fly-by-wire for re-entry, and we recommend that you do the best you can to keep a zero angle.'

Glenn throws the switch that will put him on the double manual system. 'Seven, this is the Cape, we recommend . . .' but the communication fades. The capsule is now in radio blackout.

To obtain the maximum accuracy, the capsule must revolve slowly during re-entry. Otherwise, it could splash down miles from the landing area. Using manual controls, Glenn starts the capsule rolling. There is a sudden snap. A stainless-steel retropack strap hangs in front of the window. 'This is Friendship Seven. I think the pack just let go.' No reply from the Cape.

Outside the window he can see an orange glow. It grows brighter and brighter as he watches. Suddenly there are flaming pieces of metal hurtling past. What can be happening? Is it the retropack? Or is it the capsule breaking up? He is pressed to the couch by deceleration. At any moment, he expects to

feel the heat against his back. But it doesn't happen. Peering through the window is like looking out of the center of a fireball. The brilliant orange glow engulfs the capsule.

Now back in radio contact, the Cape is calling to see how he has fared. He assures them he is 'pretty good.' The ship is back in the atmosphere, 45,000 feet (13,700 m)—manual fuel is gone. Maybe the automatic fuel, too. He reaches to deploy the drogue chute and realizes for the first time that he is very warm.

The heat has grown uncomfortable within the capsule. Despite the double-walled construction, the heat of re-entry has finally soaked through. Glenn is now waiting for the main parachute to open. The drogue pulls away and behind it appears the red and white streamer that is the main chute. The canopy fills and he feels its jerk. The capsule swings slightly on its line. The rate of descent has dropped to about 42 feet per second (12 m per second). Time now to check in with the Destroyer Noa which is the recovery ship. 'Hello, Mercury Recovery. This is Friendship Seven. Do you receive?'

'Mercury Friendship Seven, this is Steelhead (Noa's code name). Loud and clear . . . We are heading for you now.'

Glenn makes his final preparations. Glancing over his check-list, he notes that the landing bag will have to be manually deployed since the automatic sequence had to be overridden. Radio messages from the Cape remind him as well. As he pushes the switch, he feels a bump. The bag has been released. All that worry for nothing. There must have been a fluke in the telemetry signal. He completes his check list, loosens his harness and braces himself. There is a sudden firm impact and the capsule is bobbing like a cork in a bathtub.

'Friendship Seven, this is Steelhead . . . My engines are all stopped. Coming alongside at this time.'

He can feel the capsule bobbing gently against the ship. He is being hoisted up; a pause while water drains out of the landing bag. The capsule moves slowly upward, rocking with the motion of the ship. He is swung over onto the deck. The heat is intense. He listens for the sound of voices but cannot hear them. Via radio, he calls the ship and asks that the area around the door be cleared. With the word that the area is clear, he hits the firing pin which locks the side hatch. There is sudden noise. The hatch opens. John Glenn is back on Earth.

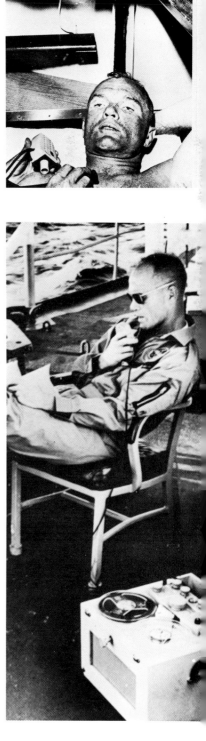

Opposite: John Glenn talks to President Kennedy by phone aboard the destroyer Noa after his recovery at sea.

Below: Glenn getting a physical checkup aboard the Noa. The microphone near his shoulder was recording his comments about the flight.

Bottom: Glenn talking into a tape recorder aboard the Noa, recording his thoughts about his historic flight

The Thrill of Victory

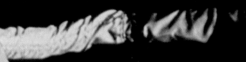

everal other Mercury flights followed John Glenn's historic three orbital flight. Fired by the marathon flights of the Russian astronauts, Nikolayev and Popovitch, NASA set out to attempt several longer orbital flights. Each flight was amazingly successful. When problems did occur, they were readily corrected without having to abort the flight. Project Mercury ended with the long flight of L Gordon Cooper.

Mercury 9 was scheduled for takeoff on 14 May 1963. But because of radar malfunctions and a problem with a diesel engine, the capsule with Cooper was not launched until the following day. Cooper made 22 orbits of the Earth before his splash-down in the Pacific. Much of his time was taken up with medical checks and photography. It was probably very boring for him, but vital for the long distance flights that were to follow.

With the end of Project Mercury, NASA began to prepare for Project Gemini. Gemini, so named because the ship would carry two astronauts, did not require all the preliminary tests that preceeded Project Mercury. There was no need for the animal flights or manned sub-orbital tests. It was, however, necessary to make some changes in the spacecraft. (By the end of Project Mercury, the term capsule was dropped.)

Gemini's spacecraft was in many ways like a larger version of the Mercury capsule. There were two seats for the astronauts arranged side by side and facing the same direction. There were four retrorockets instead of three. The interior of the craft was somewhat roomier. And the parachute which would slow the capsule for splash-down was necessarily larger. The atmosphere of the cabin which had been pure oxygen at a pressure of just over five pounds per square inch (15 kg per square cm) in the Mercury capsule, remained the same for Gemini. Because two men would be aboard the spacecraft, one had to be in command. Usually the command pilot was an astronaut who had flown in space before.

The first two Gemini flights were unmanned and were called simply GT-I and GT-II (for Gemini-Titan 1 and 2). The third spacecraft had an unofficial name, Molly Brown. GT-III, or Molly Brown, was flown

Previous spread: An excellent example of the Gemini XII space photography by Astronaut James A Lovell Jr. Astronaut Edwin E Aldrin Jr is shown during his extravehicular activities.

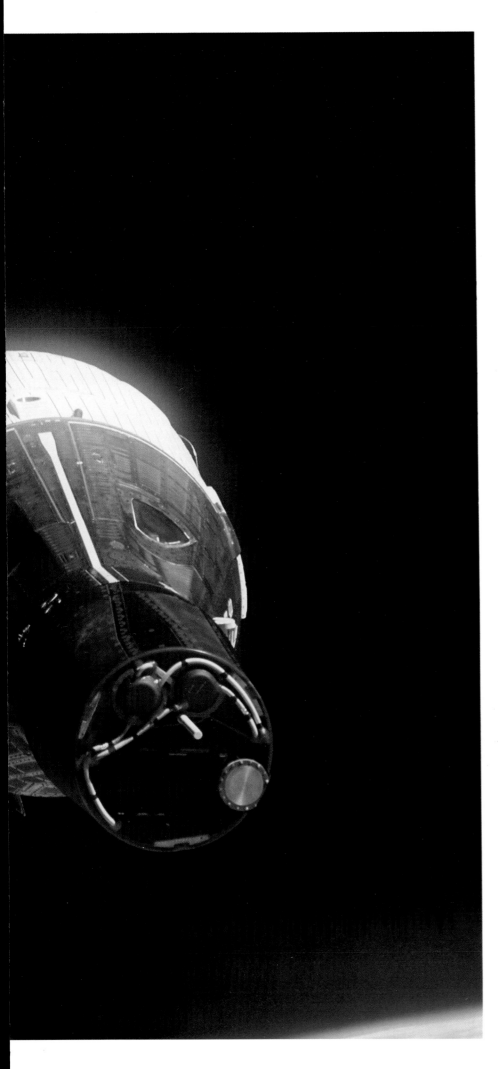

by Gus Grissom and John Young. Grissom, who was the second man in space in the Mercury suborbital flights, had had bad luck with his capsule, the Liberty Bell 7. The Liberty Bell sank at splash-down and Grissom had to swim to safety. At the time of the Gemini-III flight, there was a popular musical show on Broadway, called *The Unsinkable Molly Brown*. Hence, Grissom suggested the name to reluctant NASA officials. When he told them his second choice was the 'Titanic,' they agreed to 'Molly Brown.'

The 'Molly Brown' was scheduled for three orbits. But during these three orbits, Grissom and Young had several experiments to perform. They wanted to prove the maneuverability of the spacecraft. To do so, they changed the orbit of the craft and the inclination of the orbit. All tests went exceptionally well. Young had brought a corned beef sandwich on board which Grissom proceeded to eat, though he was scheduled to act as a control and not eat anything! Young, of course, had to stick with the special space food NASA had prescribed.

While the 'Molly Brown' remained true to her name at splash-down, the craft was some 60 miles (96 km) short of her projected landing point. To insure that the heavier Gemini craft would remain afloat, frogmen attached a special 'flotation collar' to the spacecraft before the hatches were opened. The astronauts were then lifted out by helicopter and brought aboard the carrier, *Intrepid*. Less than two hours after splash-down, the two astronauts were safely aboard.

There had been many reports that the Gemini Program was 12 to 18 months behind schedule. Many regarded this as a sign that the program would not go well. While it was true that not everything functioned perfectly on every flight, one of the outstanding successes was that the astronauts were able to override the problems. When the instruments malfunctioned, the men were able to take over.

One of the most exciting flights of the Gemini Program was launched on 3 June 1965. GT-IV was to be the first long-duration mission in the program. The astronauts for the flight were James McDivitt, command pilot, and Edward White, co-pilot. The flight

Left: The rendezvous of the Gemini-Titan VII with the GT-VI. The VII has its nose toward the camera. This was the first docking of one space vehicle with another.

was scheduled to last for four days with several interesting experiments planned for the astronauts. Most incredible was to be a walk in space for co-pilot White. Also planned was a rendezvous with the second stage Titan II rocket which would be orbiting somewhere near their craft. Since this would be a long flight, the astronauts would have to arrange sleeping periods. NASA had scheduled these so that while one man slept, the other would be awake. This did not work out well, as the pilot on duty invariably woke the one who was asleep. Future Gemini flights scheduled both men to sleep at the same time.

The rendezvous with the Titan rocket was not possible, since the expended rocket was orbiting two or three miles (3.2 to 4.8 km) distant from the ship. The walk in space, however, was dramatically successful. One hundred miles (160 km) above the surface of the Earth, Major Edward White slipped out of the spacecraft. Tethered with a gold-plated 'umbilical cord' and using a guidance gun to propel himsclf, White walked in space for 21 minutes.

When he talked about it later, White said he had 'little sensation of speed,' although the craft was orbiting at a speed of 17,500 miles per hour (28,000 km per hour). There was also no sensation of falling, for at that altitude, White was weightless. Far over the Pacific was the rosy glow of the rising Sun warming the blue and white ball of the Earth. The sight was unforgettable. The legend of the seven league boots had come true. White, in his 21 minutes in space had traveled more than 6000 miles (9600 km) . . . from Hawaii, across California, Texas, Florida and finally Bermuda.

Quipped McDivitt, who photographed the historic event, 'I wanted to make sure I didn't leave the lens cap on. I knew I might just as well not come back if I did.'

The 27-foot (8 m) umbilical served to supply oxygen and voice communication to the roving astronaut. Protected by his pressurized suit against conditions that would cause his blood to boil, White found the walk in space exhilarating. 'I just wish I had had more oxygen,' he said wistfully when the walk was over.

Above left: Commander Charles Conrad Jr.
Left: Wilcox Dry Lake in Arizona and New Mexico.
Opposite top: Astronaut Scott Carpenter looks inside the Aurora 7 spacecraft prior to insertion.
Opposite bottom: A space view of Cape Kennedy, Florida, looking southward.

Four days later, two weary but ebullient astronauts came aboard the recovery carrier, *Wasp*. They had just completed a 66 orbit, 1,700,000 mile (2,735,000 km) flight. McDivitt said of the walk in space, 'I was the happiest man in the world that day, except for possibly Ed.'

Smiling, White admitted, 'I felt so good I didn't know whether to hop, skip, jump or walk on my hands.'

GT-V was to be launched in late August of 1965. This flight was of particular interest because it was the first time fuel cells would be tested in space.

The fuel cell was a new development and scientists were eager to know how it would function in the flight. Fuel cells had been developed as a means of saving on battery weight. Almost any type of battery is exceedingly heavy. Except on the ground, they do not produce enough electric current to justify their weight. Fuel cells, however, carry hydrogen and oxygen as super-cold liquids which are then combined, forming water. In the process, they produce electrical current. Ground tests had proved that the fuel cells would produce about twice as much electrical current per unit of weight as the most efficient battery.

L Gordon Cooper, who had made the longest of the Mercury flights, was chosen as command pilot for Gemini V. The co-pilot was Charles P Conrad. The launch date was scheduled for 19 August 1965. The countdown was plagued with a number of minor problems. These caused holds that totaled nearly three hours. Countdown was finally nearing the end when a thunderstorm swept the Cape. A surge of electrical power over the launch pad knocked out the computer on board the Gemini and scrambled its memory. No launch took place that day. Two days later, GT-V lifted off in a flawless launch. A particularly lucky break in this launch was that the first stage Titan II rocket fell into the Atlantic and was recovered for study.

It had been planned for GT-V to rendezvous with an orbiting object, but nothing suitable was within reach. So for the purpose of rendezvous, the Gemini V carried a 'radar evaluation pod,' (REP) to be ejected and then later used as a target. REP was ejected from the spacecraft as planned, two and a half hours after liftoff and into the second orbit. GT-V was having problems with the new fuel cell. Conrad had reported even before the ejection that something seemed to be amiss. Pressure in the oxygen tank should have been reading at between 800 and 900 pounds per square inch (56 and 63 kg per square cm). Instead, it read at more than 150 pounds per square inch (10 kg per square cm) below normal. By the

Right: A spaceman's eye view of Southeast Egypt, showing the Red Sea.
Below: Astronaut Donald K Slayton being pressurized in a suitcheck prior to a simulated flight.

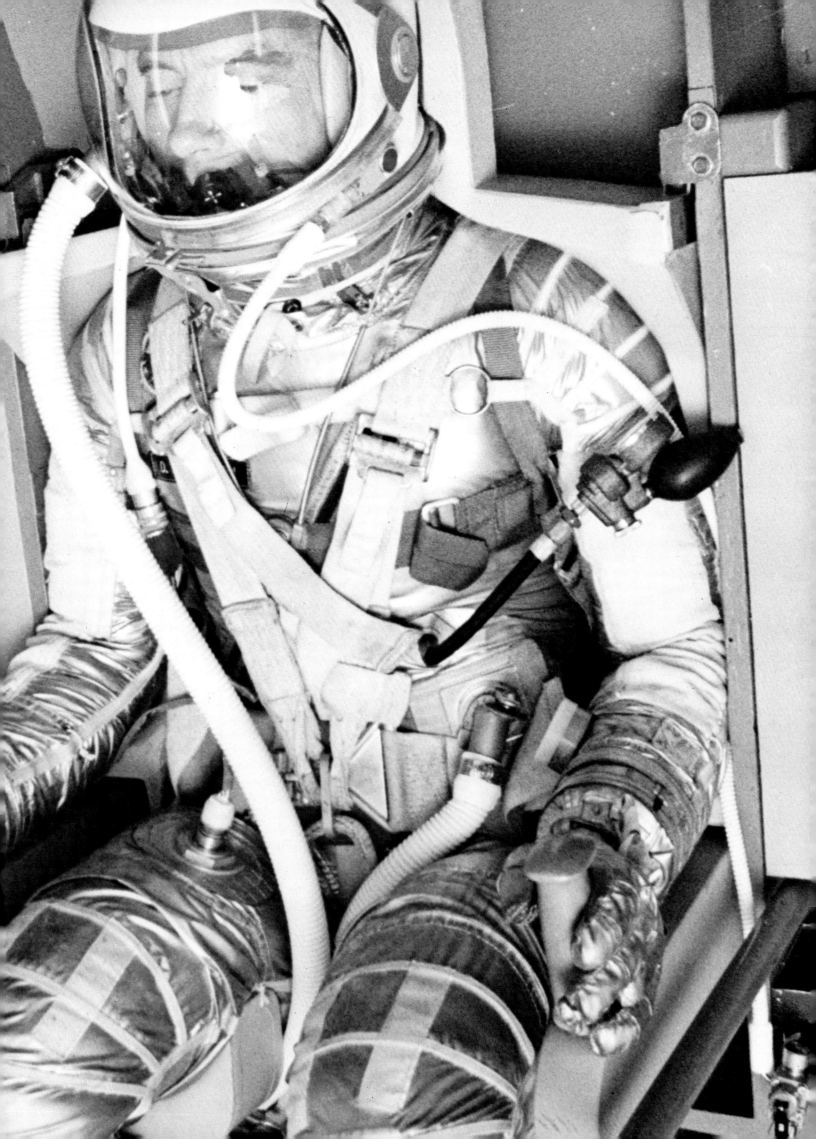

end of the second orbit, the pressure had dropped to 180 pounds per square inch (13 kg per square cm). What had apparently happened was this: the liquid oxygen had to be heated to produce the necessary pressure. The heating element for this was an ordinary glow wire, the kind found in an electric heater. It seemed that the wire was not glowing and therefore the contact was probably broken. Pressure was dropping rapidly and something would have to be done. Cooper reported to flight director Chris Kraft that he 'had to re-enter early or else power down.' Kraft ordered the astronauts to forget about the rendezvous and to shut down as many of the current-consuming instruments as possible. This Cooper and Conrad did. Then they maneuvered the spacecraft around, with the thought that the full sunlight on the oxygen tank might heat it enough to increase the pressure. It was a good idea, but it didn't work.

Conrad and Cooper were in a bind at this point. Their back-up batteries could handle another one and one-half orbits plus the re-entry. But their schedule called for them to splash down near Bermuda. To reach the recovery area, they would have to go for six orbits or wait until the completion of the 18th. Re-entry after any of these points would place the spacecraft at an inconvenient location for recovery. By the time the two astronauts and the flight director had made their decision, the pressure on the oxygen tank had dropped to 71 pounds per square inch (five kg per square cm). But it seemed to have stabilized at that point. On the basis of this information, Kraft gave permission for a GO for 18 orbits. Oddly enough, on the continued flight, the situation improved because the spacecraft itself was producing heat, which the fuel cell tank absorbed.

Now that it was apparent that the flight would continue, the astronauts began a series of experiments to test their vision of objects on the ground. Most of the experts tended to disbelieve the reports of earlier astronauts. So for GT-V, special 'eye charts' were laid out in New Mexico and Australia. Cooper and Conrad had no trouble at all in recognizing the signs, even those which they were not supposed to recognize! So excellent did their vision prove to be that Conrad reported to ground control the presence of a carrier and a destroyer at the harbor entrance of Jacksonville. It turned out that the 'carrier' was actually a tugboat and the 'destroyer' a lighter. Such vision at 100 miles (160 km) above the earth seemed rather unbelievable.

Despite the problems they incurred, Cooper and Conrad flew a nearly complete mission of 120 orbits. They were scheduled

to make 121, but splashed down a little earlier because of a hurricane approaching the re-entry area. Their flight turned out to be one of the quietest, despite all the early anxieties.

NASA was still interested in a space rendezvous. GT-VI was planned as the first flight that would make an actual rendezvous in space. The rendezvous target was to be an Atlas-Agena rocket which would be launched 101 minutes before the GT-VI. Countdown for the launch was begun on 25 October 1965. The two rockets, the Titan II with the GT-VI and the Atlas bearing the Agena, were counted down simultaneously. The Atlas roared off as planned. But there was no telemetry from the Agena. Ground control despaired of the mission, but waited to hear from the Australian tracking station. It was possible that the Agena had only suffered a trans-mitter failure. The Australian station would be able to pick up the Agena via radar echo. But a few minutes before noon, the message from Australia told it all, 'no joy, no joy.' Since there was no rendezvous target, countdown for the GT-VI was scrubbed.

The flight of GT-VI would have to be postponed indefinitely, since there was no other Agena available. Meanwhile, GT-VII had passed all preliminary check-outs and was launched on 4 December 1965. The command pilot was Frank Borman and the co-pilot was James Lovell. They were scheduled to make a two-week long flight. At this point, a bright idea came to officials at NASA. With the successful launch and orbit of GT-VII, there was a possible rendez-vous target for GT-VI! Take-off for GT-VI was 12 December 1965. But the launch was plagued with problems and was scrubbed until 15 December. On the morn-ing of 15 December, GT-VI was launched into orbit. The command pilot was Walter Shirra and the co-pilot was Thomas P Stafford. The early orbit of the craft showed a perigee of 100 miles (160 km) and an apogee of 161 miles (258 km). The GT-VII was orbiting at a distance of 185 miles (296 km). Using a series of short burns, Shirra and Stafford maneuvered their ship into an orbit of 185 miles (296 km). Approximately six hours after lift-off, the two ships were within one and one-half miles (two and one-half km) of each other. Fifteen minutes later, they were within 20 feet (6 m) of each other. Photographs taken showed that the two spacecraft had come within two or three feet (.5 and .9 m) of each other. Rendezvous had been accomplished with amazing success. The two spacecraft remained in orbit for some five and one-half hours after which Schirra and Stafford splashed down on the morning of 16 December. GT-VII completed its scheduled two weeks in orbit and splashed down on 18 December. Both Borman and Lovell were in excellent condition. They thought this was partly due to the fact that they had flown for much of the time without their spacesuits.

Gemini VIII was scheduled for lift-off on 16 March 1966. Neil Armstrong would act as command pilot and David R Scott would be co-pilot. Their flight plan contained not only a rendezvous, but an actual docking exercise. The Gemini spacecraft would dock with an Agena rocket. The simultaneous countdown began. The Atlas-Agena launched without a hitch and was soon orbiting the earth in a circular orbit of 168 miles (269 km). The GT-VIII followed with its first orbit running between 100 and 168 miles (160 and 270 km) from the Earth's surface. Gradually, the GT-VIII was brought into a circular orbit of 167 miles (247 km).

All was going very smoothly. Docking was accomplished readily during the fifth orbit over the Brazilian coast. Gemini VIII approached and docked with the Agena rocket at a speed of one and one-half miles (two and a half km) per hour. NASA had feared an electrical discharge might take place between the two ships, but none occurred. There was scarcely a jolt as docking took place. The astronauts, carefully turning the Gemini, satisfied themselves that the two vehicles were locked together. The orbit continued uneventfully for the next 40 minutes. Then, without warning, the two craft began to pitch and roll violently. Armstrong and Scott had no idea what was causing this behavior. They were fearful that the two ships would come apart. In such a break-up, the GT-VIII might be disabled. Another problem was the fact that the fuel tanks of the Agena were still more than half full. It was possible that they might split open and explode. Very quickly, Armstrong and Scott separated themselves

Opposite top: A photograph of NASA's Gemini VII taken through the hatch window of the Gemini VI during rendezvous, 15 December 1965.
Opposite bottom: Side view from 38 feet of the augmented target docking adaptor (ADTA).
Top right: Docking with the Agena near the Philippines.
Center right: Commander Cernan's EVA camera.
Right: A rear view of the ATDA from 47 feet.

45

from the Agena. Flight Control suggested an emergency landing. The astronauts, deeply concerned by the unforeseen problem, decided to go for the emergency landing. They splashed down 550 miles (880 km) from Okinawa, only about a mile from their estimated landing point. Splashdown had gone perfectly. Armstrong and Scott were rescued in a short time by the destroyer *Mason*.

It was later discovered that the problem had been a short circuit in one of the thrusters. Had the astronauts known the source of the problem, they might have continued their flight for the full 71 hours originally scheduled. When it was realized that nothing was wrong with the Agena, NASA ignited its rocket motors by radio command. The Agena was thus brought into a 253 mile (405 km) orbit where it would serve as a docking target for a later flight.

GT-IX was scheduled as a repetition of the tests performed by GT-VIII, plus a few others which Armstrong and Scott had not had time to carry out. Launch date for Gemini 9 was scheduled for 17 May 1966. As in the past, Gemini would be preceded by

an Atlas Agena rocket launch. The Agena, unfortunately, had a problem.

An airline pilot, flying through the area at the time of the launch, described it graphically: 'We first spotted the rocket's vapor trail at about 15,000 feet (4570 m) as it broke through a cloud layer. We watched it rise vertically in front of us and suddenly there was a great yellow eruption and we knew something unusual had happened. An object fell from the base of the yellow cloud and dropped toward the ocean.'

What had happened was that a short circuit in the Atlas caused one of the booster engines to make a sharp loop. The Atlas broke up and the Agena fell, unignited, into the sea. There was no other Agena rocket available. None would be ready for some time. But, since GT-IX was ready, NASA decided to go ahead and try another device called an ATDA or augmented target docking adaptor. The ATDA had no propulsion system of its own. It needed a conical shroud to serve as a nose cone and an Atlas rocket to carry it into orbit. ATDA was launched on 1 June 1966. NASA waited to launch Gemini-IX until they were sure the ATDA was in orbit. When the tracking stations

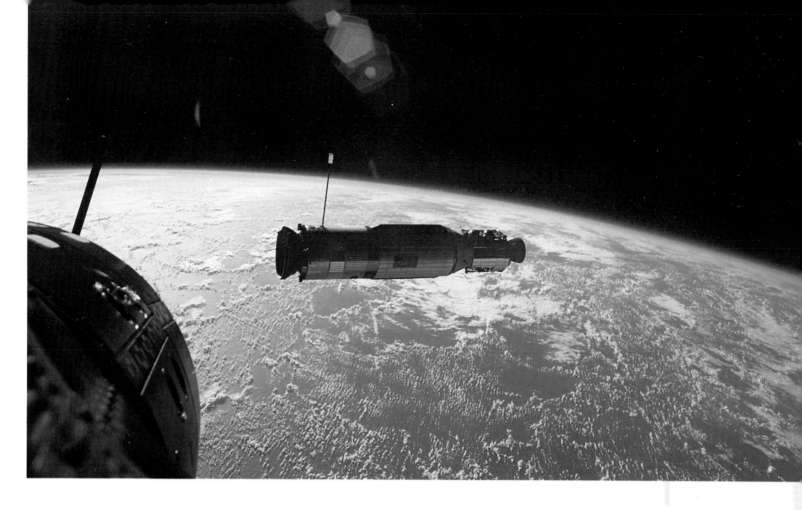

reported they had picked up the ATDA, the GT-IX was launched. But when Thomas Stafford and Eugene Cernan approached the ATDA, they saw a very odd sight. Stafford reported back to the ground crews, 'We have an angry alligator on our hands.' The conical shroud on the ATDA had only split open lengthwise and was still attached to the target. (Ground crews had never handled an ATDA before and they had made a mistake in the wiring.) The split nosecone looked exactly like the jaws of an alligator. Docking was going to be impossible. Should Gemini try to knock the shroud loose? Ground control voted 'No Go' fearing that to do so might damage the ship. Gemini IX made three different rendezvous maneuvers, instead.

Eugene Cernan was to perform some tasks outside the spaceship. It was necessary for him to work his way around to the back of the ship and retrieve a backpack called an AMU (astronaut maneuvering unit). Cernan found that working in a weightless condition was very difficult. Often he had to stop and rest. The face plate on his helmet kept fogging so much that he could scarcely see. Finally after two hours outside the confines of the ship, Stafford ordered him to come back inside.

Tasks for the Gemini astronauts grew more complex as the program progressed.

GT-X, launched 18 July 1966, was also destined for a space rendezvous. This time it would be a little more spectacular. Serving as command pilot was John Young and the co-pilot was Michael Collins. Timing for their launch was of the utmost importance. They took off exactly 101 minutes after the Agena they were to dock with lifted off. They had only a 40 second leeway. Had they missed it, there would have been a two-day delay in their launch. GT-X made contact with the Agena in late evening. For nearly 40 minutes, the astronauts followed the orbiting Agena before actually docking with the rocket. This procedure used up more fuel than they had anticipated. They had expected to have 770 pounds (350 kg) left after docking but their supply had been reduced to 385 pounds (175 kg). What to do? Docking had been successfully accomplished. The Agena had fuel left in its tanks. Young checked the orbiting rocket and then fired its motor from his control panel. The remaining fuel in the Agena served to carry the joined spacecraft into an orbit with a perigee of 184 miles (295 km) and an apogee of 474 miles (758 km)—a record for the highest manned flight at that time.

Above: Agena tethered to Gemini over the Gulf of California near La Paz, Mexico.
Opposite: Gemini docked to the Agena, with the hatch open.
Below: A space view of the coastal areas of Guyana and Surinam.

Young and Collins now had another mission to perform. They were to adjust their orbit to coincide with that of Agena 8, the rocket that had been left in orbit by the Gemini VIII flight. The Agena 8 was, at that point, some 1400 miles (2240 km) distant. The astronauts were able to see it as a very bright object ahead of their ship. The GT-X had been docked the Agena 10 for 39 hours and 40 minutes, when on the afternoon of 20 July, the two craft were finally separated. One hundred and sixty miles (256 km) ahead and seven miles (11 km) farther out was the Agena 8. Within three hours, Young and Collins had bridged the gap between the two vehicles and were orbiting alongside the Agena 8. Fuel was low and there would be no docking exercise. Collins, however, went outside the ship and strapped himself into the backpack unit. On his third attempt, he was able to reach the Agena-8. After 27 minutes outside the ship, Michael Collins re-entered the GT-X.

With the splashdown of the GT-X, the Gemini program had only two more flights to go. The remaining two flights would try to refine the accomplishments of the preceeding flights. GT-XI lifted off the Cape on 12 September 1966. Her command pilot

face and blinded his vision in his right eye. Conrad ordered him back into the ship after about 10 minutes. But he had been successful in attaching the 100 foot (30 m) tether. They could begin their experiment with artificial gravity.

Using the thrusters of the Gemini, the ships were set to spinning at the rate of one rotation every nine minutes. The gravity generated didn't amount to much. The astronauts were unable to feel anything. But four other separate docking maneuvers were successfully performed before splashdown took place.

GT-XI was the first spacecraft to re-enter the Earth's atmosphere using computer control only. Previously, astronauts had always had overriding control. The splashdown was picked up by TV cameras stationed on the aircraft carrier *Guam* and carried by the Early Bird satellite. The entire recovery took place in only 25 minutes.

GT-XII was scheduled to be the last flight of the Gemini Program. It was launched on 11 November 1966, preceeded by an Atlas-Agena. The command pilot for the GT-XII was James A Lovell and the co-pilot was Edwin Eugene Aldrin. Lovell had made the two week marathon flight with Frank Borman on the GT-VII. Aldrin was an Air Force major with a PhD in science from MIT. His doctoral thesis had been on orbital mechanics and rendezvous maneuvers. It was, however, his flying skills that were to prove of great value in this particular trip.

One of the objects of the flight was to rendezvous with the Agena. But a computer malfunction almost scotched that procedure. Aldrin noted early on in the flight that the computer wasn't registering the change in distance to their target. The two astronauts were unable to do anything about the problem, but decided to try to make the rendezvous using manual controls and visual sightings. This proved very successful.

was Charles P Conrad and her co-pilot was Richard F Gordon. Their mission was to rendezvous and dock with the Agena 11 on the first orbit. To do this required a precision launch of both the rocket and the Gemini-XI Titan rocket. There were minor problems in the launch that caused a day's delay.

Conrad and Gordon were able to dock with the Agena 94 minutes after their launch. They remained docked for several hours. Then, using the Agena's fuel, Conrad put the joined craft into a new orbit. The perigee for this orbit was merely 175 miles (280 km) but the apogee was 850 miles (1360 km), a height that outdid the previous GT-X flight spectacularly. They then returned to an orbit of 180–190 miles (290–304 km) above sea level.

The next experiment that Conrad and Gordon would perform was to try to create an 'artificial gravity.' They released the GT-XI from the Agena, but kept the two vessels tethered together. The two ships would then be rotated around their common center of gravity. The force thus created in the Gemini spacecraft would act as gravity. The task of attaching the tether fell to co-pilot Richard Gordon. Gordon found, as Eugene Cernan had, that movement in the weightless environment outside the craft was extremely tiring. Perspiration ran down his

Above: Major Edwin E Aldrin Jr's EVA exercise.
Opposite top left: A space view of the Nile Valley from Girga to the Delta, the Gulf of Suez and the Sinai Peninsula.
Opposite top right: Looking north, a space view of the Maldive Islands with the Arabian Sea to the left and the Bay of Bengal to the right. The altitude is about 400 miles.
Opposite bottom: Major Edwin E 'Buzz' Aldrin inside his spacecraft.
Below: A photo showing the Agena on tether. In the background are clouds over the Pacific Ocean.

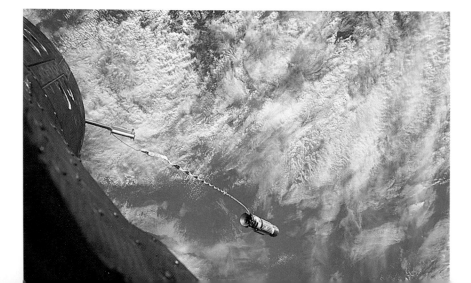

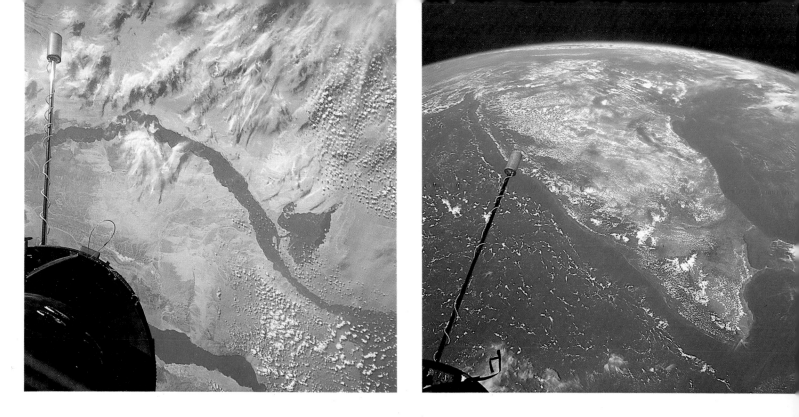

Aldrin was scheduled to spend an extended period of time outside the spacecraft. He had been concerned about the exhaustion which had been such a problem for other astronauts. After careful thought, he felt he knew how to deal with the problem. He slowed all his movements to a snail's pace. This worked amazingly well. Aldrin was able to perform his tasks with none of the overwhelming weariness that struck Cernan and the others. Aldrin spent a total of five and one-half hours in space, unscrewing bolts and tightening them, checking electrical connections and other things. When it was time to re-enter the spacecraft, he carefully wiped the windshield. From inside the craft, Lovell called, 'Check the oil, too.'

Astronomers back on earth were particularly interested in the photographs made on these missions. Photos from space avoided all the distortions prevalent in the Earth's atmosphere. Of particular interest were the 'hot' stars and the pictures of an eclipse taken from space. Scientists from other countries were eager to get photos of some of their high altitude rockets which Aldrin had taken during the flight. After 59 orbits, the Gemini-XII returned to Earth.

With the flight of the GT-XII, the Gemini program was over. The score was even—ten manned flights—ten successes. Much valuable knowledge had been gained. What had been of particular importance in the Gemini program was the resourcefulness of the astronauts themselves. Even on the threshold of space, Man was still in control.

Disaster and Triumph

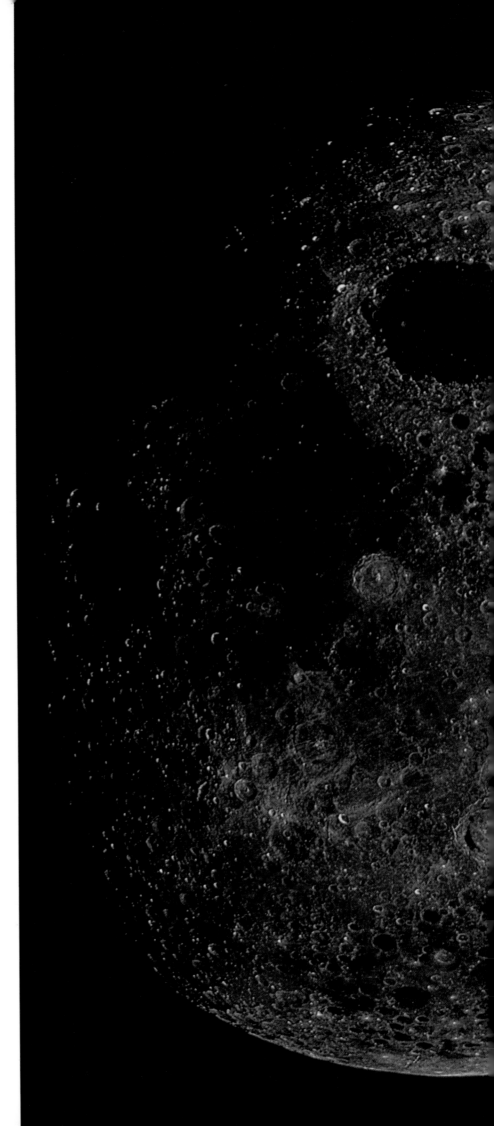

All during the Gemini program, NASA had also been preparing for Project Apollo whose ultimate goal was to place a man on the moon. An unmanned craft, Surveyor I, had already been landed on the moon and lunar orbiters had sent back fantastic photographs of the moon's surface. Precise testing of the components for the Apollo spacecraft, command module and the LEM (Lunar Excursion Module) were all underway long before the conclusion of the Gemini program.

The Gemini program had been one of great success. It was, therefore, hard for most laymen to understand that there was considerable risk and danger involved. 'Gus' Grissom had once addressed this question in a press conference. He said, 'If we die, we want people to accept it. We are in a risky business, and we hope if anything happens to us it will not delay the program. The conquest of space is worth the risk of life.' All through history, explorers, whether they crossed vast oceans in small boats, climbed the highest mountain tops or ventured into the depths of unexplored continents, have faced death in seeking knowledge.

The first Apollo command module was scheduled for launch on 21 February 1967. The crew of that first flight was to be 'Gus' Grissom (the second American in space), Edward H White (the first American to walk in space) and Roger B Chaffee (who would be making his first flight above the atmosphere). On 27 January 1967, the three were scheduled for some routine exercises aboard the command module as it was perched above the Saturn B-1 rocket 220 feet (67 m) above the ground. The spacecraft, like all the previous ones, was pressurized to 16 pounds per square inch (1 kg per square cm) with pure oxygen. The astronauts wore suits pressurized with less pure oxygen. It was early evening, about 6:30 EDT when a voice called out, 'Fire in the spacecraft!' Another voice cried, 'Get us out of here!' Technicians on the gantry saw a sheet of flame inside the module. Wearing face masks and asbestos gloves, they tried valiantly to open the hatch. But they were driven back by the intense heat and smoke coming from the capsule. Some six minutes

Right: A photograph of a nearly full moon taken from the Apollo 8 spacecraft—29 December 1968. *Previous spread:* This view of the rising Earth greeted the Apollo 8 astronauts as they came from behind the moon after the lunar orbit insertion burn. On the Earth, 240,000 statute miles away, the sunset terminator bisects Africa.

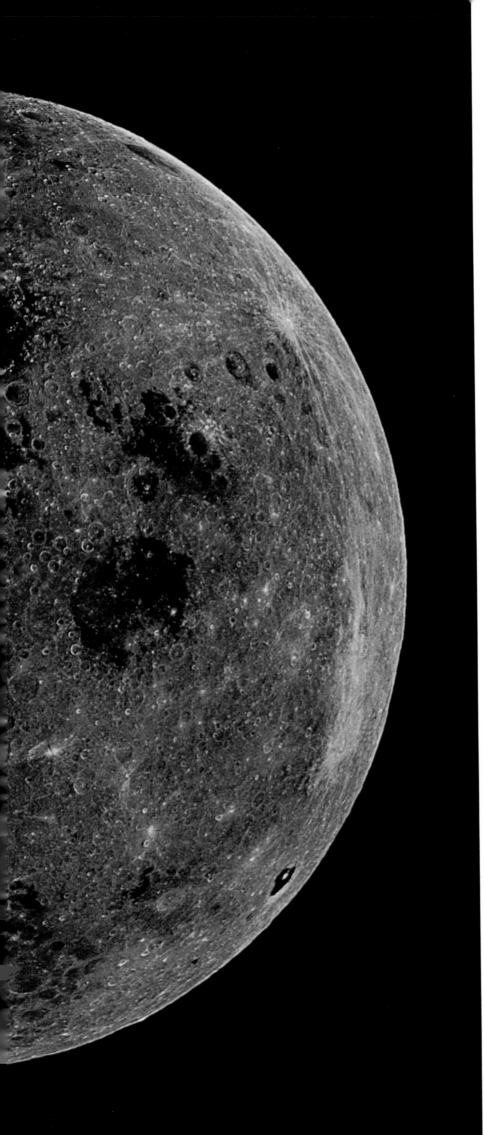

after that first alarm, they were able to remove the hatch. It was too late, however, for the three astronauts. They had died almost instantly in the smoke and flames that destroyed the capsule. The accident left the nation speechless with shock. These were the first astronauts to lose their lives in the line of duty.

There were, of course, inquiries into the matter. The capsule was so badly burned that it was almost impossible to determine what had actually caused the fire. It was believed that it probably had started with a short circuit. Had it been possible to open the hatch instantly, the astronauts might have been able to escape, protected temporarily by their pressurized suits. The atmosphere of pure oxygen within the capsule was a great fire hazard. But it had been used with so much success in all the preceding flights that no one had worried much about the possibility of fire. After 'Apollo 204,' as the accident was referred to, normal atmosphere within the capsule was used for all ground tests. Pure oxygen at reduced pressure was used after orbit had been achieved.

For a while there was silence at the Cape. But the words of 'Gus' Grissom were a challenge to the remaining astronauts. 'We are in a risky business, and we hope if anything happens to us it will not delay the program.' And so Apollo went on as scheduled. There was anxiety at first. There were greater precautions than before. But it seemed the only fitting tribute to three brave men who had given their lives for a chance to reach the moon.

For the moon was, indeed, the goal of the Apollo program. Less than two years after the tragedy of Apollo 204, Apollo 8 was launched in December 1968. The flight was one of the most awe-inspiring of the entire space program. Three men, Frank Borman, command pilot, James Lovell, navigator, and William Anders, systems engineer, were carried into an orbit around the moon. At one point in this orbit, they were only 70 miles (112 km) from the surface.

The historic flight had begun on the morning of 21 December 1968. At exactly 2:36 am the three astronauts were awakened. Dressed in their white spacesuits, they took the elevator 320 feet (100 m) upward to the command module. Time seemed to crawl as the crew and ground control went through endless checklists. At 5:34 am (T-minus 2 hours and 17 minutes), the command module's hatch was closed. All was going perfectly. The count-down was proceeding without a hitch and tension was mounting among those who sat watching the screens at Houston's Manned Spacecraft Center. T-minus 9 seconds, ignition, and liftoff into a balmy December sky; the 6.2-million-pound (2.8 million kg) vehicle which burned 15 tons of fuel per second had cleared the tower. Its destination lay 230,000 miles (368,000 km) away. The roar of the booster rocket, a Saturn V, was so deafening that ground control could scarcely hear the reports from the astronauts. Curiously, onlookers watching the liftoff some three miles (4.8 km) away, saw flames shooting hundreds of feet in the air, saw the rocket rising skyward, but heard no sound. There was only an eerie silence. Then, fifteen seconds later, there was an earth-shaking roar as the rocket streaked toward space. For $11\frac{1}{2}$ minutes, the craft climbed with ever-increasing speed until it reached a speed of 17,428 miles (27,000 km) per hour.

Meanwhile, the men in the command module surveyed their display panel closely. On its surface resided 24 instruments, 40 event indicators, some 71 lights, and 566 switches. In these first minutes, Borman sat with his hand on the controls, ready to take action if the need arose. But the automatic control system operated with the precision of a fine Swiss watch. The first and second

stages fell away into the Atlantic.

At this point, Lovell unbuckled himself from the center couch. As navigator, he had an assignment to do. But suddenly, he found himself floating as he put it, 'all over the place.' In previous flights, he had always been strapped down and the capsule had offered little room for floating. The weightlessness made him feel a little seasick. But he slowed his movements and the feeling of queasiness went away. Carefully, he made his first sightings using two optical instruments: a scanning telescope and that age-old instrument, the sextant. With these he made readings on the angles between certain stars and the horizon of either the Earth or the moon. These enabled him to check the spacecraft's gyroscope and locate the craft in space as well as confirm its speed and direction of travel.

For nearly three hours, the command module remained in orbit around the Earth. During this time, both ground control and the astronauts were actively checking everything—the launch vehicle, the command module, and the ground stations—for the slightest hint of a malfunction. This was what NASA officials referred to as a 'Go/No Go' situation when the actual decision to go to the moon would be made. The flight was proceeding with textbook perfection. Then, as the spacecraft came within sight of Hawaii, the third-stage engine was fired. It

Above: An Apollo 8 moon view, photographed at a distance of 150 nautical miles as the Apollo 8 spacecraft orbited the moon on 24 December 1968. The lunar surface has less pronounced color than indicated by this picture.

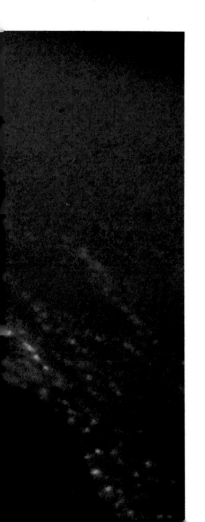

Below: A pod capsule camera mounted on the second stage of NASA's Saturn V launch vehicle photographed the ignition of the five J-2 engines and the separation of the first and second stages (S-IC and S-II).

would produce one of the longest burns of the mission, lasting more than five minutes. It was what NASA called the TLI or translunar injection burn which would put the spacecraft on a pathway around the moon and back to Earth. Once the burn ended, the men in the command module were traveling faster than human beings had ever gone before—24,226 miles per hour (39,000 km per hour)! From the Earth, far below, came the voice of Chris Kraft, Director of Flight Operations. 'You're on your way. You're really on your way.'

Trajectory was A-OK. Borman now performed the first manual operation of the flight. Pulling a T-shaped handle, he triggered the explosive devices that separated the third stage rocket from the spacecraft. The shock of the explosion was somewhat rougher than the crew had anticipated, but the separation was complete.

The third stage, however, trailed the spacecraft a little too closely. It seemed to be only about 500 to 1000 feet (150 to 300 m) away. Unused fuel was spewing out from all sides of the abandoned rocket. Droplets of the fuel were so thick, they resembled a snowstorm or as Lovell put it, 'millions of stars.' A further short burn of the reaction control engines put distance between the spacecraft and the rocket.

One of the things which the astronauts checked on this historic flight was the amount of radiation picked up as they passed through the Van Allen Radiation Belt. Each of the astronauts wore a dosimeter for this purpose. As ground control had expected, the readings were negligible.

At 21,000 miles (34,000 km) from the earth, James Lovell gave this description of the view: 'I'm looking out my center window, the round window, and the window is bigger than the Earth is right now. I can clearly see the terminator [the line which separates the daylight from darkness]. I can see most of South America all the way up to Central America, Yucatan and the peninsula of Florida.' The view was, as usual from outer space, amazingly clear. However, the center window had begun to fog up. This proved a problem for the remainder of the flight. What was happening was that the sealing compound in the windows was causing a deposit between the triple-layered windows. It was a minor, if rather frustrating, problem.

The velocity of the spacecraft had begun to slow noticably. At 25,900 miles (41,000 km) out from the earth, the speed had dropped to a mere 8600 miles per hour (14,000 km per hour). Laboring against the gravitational pull of the Earth, the small ship continued to steer a course for an orbit around the moon.

An interesting maneuver at this point was known as the 'barbecue mode' in the parlance of ground control. Officially it was called PTC or passive thermal control. At this time, the spacecraft would roll slowly on its long axis which faced the Sun. This slow roll would occur once every hour to insure that the craft was evenly exposed to the heat of the sun.

Another maneuver had to be performed every ten hours or so. This was the oxygen fuel-cell purge. Within the service module, which was the portion of the craft just behind the command module, were three fuel cells. These generated electrical power for the Apollo. Within the cells, hydrogen and oxygen reacted, producing electricity, drinkable water, and heat. Hydrogen and oxygen passed through a porous metal plate. To insure that the plate did not become clogged, it was flushed out about every ten hours.

Sixty thousand miles (96,000 km) into space, all systems were Go: it was now time for another critical maneuver. This was another major burn which would effect a midcourse correction. It would make use of the service propulsion system (SPS). This particular system had not yet been tried on this trip. The engine was known to be extremely reliable, but its performance now would be critical. The SPS was the spacecraft's only significant means of propulsion. The astronauts would rely on it to put them into lunar orbit and also to send them out of lunar orbit and back to Earth. Would it function as reliably as in the past? Mission Control in Houston watched its screens anxiously. Pressure and temperature readings were normal, which indicated the engine was working well.

The CAPCOM calling in from Houston reminded Borman that it was now time for him to 'hit the sack.' The commander was more than ready to turn in for a little much-needed rest. But, whether due to the excitement of the trip or over-tiredness, he found he was unable to sleep. He requested permission to take a sleeping pill, which ground control granted. A simple enough request, but the action was to cause some anxious moments later.

Lovell and Anders remained awake, taking care of the housekeeping chores.

Above: A striking view from the Apollo spacecraft (left) showing nearly the entire Western Hemisphere from the mouth of the St Lawrence River, including nearby Newfoundland, extending to Tierra del Fuego at the southern tip of South America. A small portion of the bulge of West Africa is at the right.

Below: Astronaut Frank Borman, the spacecraft commander, is shown during intravehicular

Periodically, they would have to dump waste water from the fuel cells. There was a good deal of light-hearted joking between the astronauts and Mission Control as they performed these little routines.

'You're looking pretty small down there now, Houston.'

'We're carrying a big stick, though. . . .'

'OK, we're dumping now, Houston . . . we finally got some stars to see.'

This last remark was in reference to the fact that James Lovell was having trouble seeing any real stars because of light scattering in his telescope. The waste water, dumped into the chill of outer space, formed ice crystals immediately. It was this that caused a blizzard of 'stars.'

Unlike the captains of old, Frank Borman was too busy much of the time to write anything down. Instead, on board the Apollo there was a tape machine on which the astronauts recorded messages when the craft was out of communication with the ground stations. These tapes served as a kind of ship's log. From time to time, Mission Control played back these tapes to check on the condition of the men in the command module. They were therefore startled to learn from the tapes that at one point, Frank Borman had been ill and that the other crew members had been complaining of nausea. A bout of illness could have been a very serious matter on this journey. As it turned out, the illness was nothing serious. Borman felt most of it was due to the sleeping pill he had taken rather than any virus. The Apollo flight surgeon, Dr Charles Berry, spoke directly to the astronauts to determine just what their condition was. They all reported that they were feeling much better. Dr Berry suggested some medical treatment should the illness recur. At this point it would have taken nearly 34 hours to bring the astronauts back to Earth. Since all were reporting feeling much better, Dr Berry recommended that the flight continue.

Halfway to the moon, Borman, Lovell and Anders showed some TV pictures of their flight. They had brought a tiny TV camera which weighed a mere 4.5 pounds (two kg).

Using it, they showed a bit of life aboard the spacecraft. Such vignettes as James Lovell mixing up a bag of freeze-dried chocolate pudding and William Anders reaching toward a floating toothbrush brought the reality of the trip home to earthbound viewers. The object the astronauts wanted most to show was the view of the Earth, but the light was too bright for a clear picture. Most of the TV pictures that people had seen heretofore had been only simulated drawings.

In a second TV presentation, Lovell described the view from the spacecraft. 'The Earth is now passing my window. It's about as big as the end of my thumb. Waters are all sort of a royal blue; clouds, of course, are bright white. The reflection of the Earth is much greater than the moon. The land areas are generally sort of dark brownish to light brown. What I keep imagining is, if I were a traveler from another planet, what would I think about the Earth at this altitude? Whether I think it would be inhabited.' They were now 200,000 miles (320,000 km) from the Earth.

At 38,900 miles (62,000 km) from the moon, the spacecraft had begun to react to the pull of the moon's gravity. The craft, which had dropped to a speed of 2223 miles (3500 km) per hour, began to accelerate. The trajectory had been so accurate that

two of the planned midcourse corrections had been canceled. But, in ejecting the waste water from the fuel cells, the craft had picked up some velocity. To slow the speed, four of the reaction jets were engaged in a retrograde burn of 11.8 seconds. The effect was to slow the craft by about 1.4 feet (43 cm) per second and bring the closest approach to the moon to about 70 miles (112 km).

As the craft drew nearer the moon, there were other preparations to make. The spacecraft would have to be propelled into an orbit around the moon. To accomplish this, the reaction jets would have to make a LOI, lunar orbit insertion burn. Careful and repeated checks on the navigation, computer and spacecraft systems were required as well. Finally when all seemed complete, the men aboard the Apollo took a little rest. Not, however, before they gave Mission Control a rather startling announcement— 'As a matter of interest,' Frank Borman commented, 'We have yet to see the moon.'

'What are you seeing?' inquired CAPCOM.

'Nothing,' replied Bill Anders. 'It's like being on the inside of a submarine.'

Only one of the astronauts had caught even a glimpse of the moon. That was James Lovell, who had seen a small crescent through the lens of his telescope.

It was now Christmas Eve, 1968. The time was early morning. The clock on the wall at Mission Control was counting down the time until LOS, loss of signal. The loss of signal would occur at the moment the spacecraft slipped behind the moon. Tension mounted as the craft, accelerating ever more rapidly, approached the moon. The crew worked silently, too intent on their instruments to offer comment. From Mission Control came the voice of Jerry Carr. 'One minute to LOS. All systems Go. Safe journey, guys.'

At Mission Control, the seconds and minutes stretched agonizingly long. There was nothing to do but wait.

On board the spacecraft, the crew of the Apollo closely watched the display and keyboard of their computer. Information flashing across the screen was being updated with information from the ground. Thirty seconds before LOI, a final countdown flashed on the screen. At T-5 seconds, a Go/No-Go symbol flashed on the screen. The computer was waiting for final confirmation to proceed. Lovell pressed the 'proceed' key and the service propulsion engine fired.

activity on the Apollo 8 lunar orbit mission. The print was taken from movie film.

It burned for 246.9 seconds, slowing the speed of the spacecraft to 3721 miles (5960 km) an hour. The craft was now in an elliptical orbit around the moon. At its lowest point, the Apollo would be only 70 miles (112 km) from the moon's surface. At the highest point, the craft would be 190 miles (300 km) above the moon.

Some 35 minutes after LOS, the voice of James Lovell was heard at Mission Control. There was a collective sigh of relief. They had made it! The immediate concern at this point was for the condition of the tank pressures and the engine, the performance of the water evaporators that kept the spacecraft cool. All appeared to be performing well. Then came the question that all the world was asking: what does the moon look like?

The moon bore sharp contrast to the blue and cloud-shrouded Earth 230,000 miles (368,000 km) distant. As Lovell first described it, 'The moon is essentially gray, no color. Looks like plaster of Paris. Sort of grayish sand . . .'

Frank Borman's description was even more graphic. 'Vast, lonely and forbidding.'

Photos showed a gray sphere pockmarked with craters great and small, but most of them essentially round without the ragged appearance such formations would have on earth. Said William Anders of moon photography, 'All you really need is black and white film.'

The photos were amazing and impressive in their detail. Forty-mile (64 km) wide

Above: An Apollo 9 view of the docked Apollo 9 Command/Service Module 'Spider' with Earth in the background, during Astronaut David R Scott's extravehicular activities on the fourth day of Apollo 9's Earth orbital mission. Below: A view of the Apollo 10 modules.

craters lying like broken bubbles, smaller craters, radiating mysterious white lines, the vision was almost miraculous. TV audiences back on Earth were learning a whole new moon geography. There was the Sea of Tranquility, the Sea of Crises, and the Sea of Fertility, groups of craters known as Columbo and Magelhaens. It was also possible for the first time ever to get a look at the dark side of the moon. Borman described it as 'a dark, unappetizing looking place' not as crater-riddled as the visible side. None of the variations of color that are so familiar on Earth were visible on the moon's gray landscape. The astronauts, who had made a study of the moon maps back on Earth, had no trouble identifying geographic features.

In the midst of all the photography, the messages between Houston and Apollo 8, TV pictures and flight duties, the spacecraft orbited the moon functioning as smoothly as clockwork. The men, however, were growing tired. They were finding sleep difficult. Frank Borman told Houston that 'the flight plan looks a lot fuller than it did in Florida.'

Back on Earth it was Christmas Eve. Frank Borman, who was a lay reader in his church, read a prayer to be recorded and broadcast on the ground. 'Give us, O God, the vision which can see Thy love in the

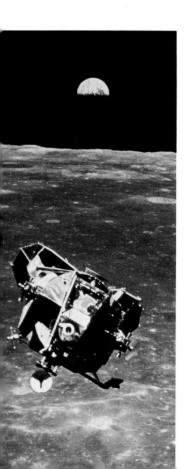

Above: Astronaut Commander Edward H White II on extravehicular activities over New Mexico.
Below: The Apollo 11 Lunar Module ascent stage photographed from the command service module during rendezvous. The LM was making its docking approach.

world in spite of human failure. Give us the faith, the trust, the goodness in spite of our ignorance and weakness. Give us the knowledge that we may continue to pray with understanding hearts, and show us what each one of us can do to set forth the coming of the day of universal peace. Amen.'

'Apollo 8. This is Houston. Your TV show was a big success. It was viewed this morning by most of the nations of your neighboring planet, the Earth. It was carried live all over Europe, including even Moscow and East Berlin. Also in Japan and all of North and Central America, and parts of South America.' This brief announcement gave only a slight indication of the international applause that accompanied the flight of Apollo 8.

But the crew was tiring. James Lovell's flight plan had kept him constantly busy; Anders had had to work in meals around his photography schedule; Frank Borman, as Commander, had to be aware of the functioning of both the crew and the spacecraft. It would be vital for everyone to be alert for the transearth injection or TEI. If it did not succeed, the craft and her crewmen would remain in orbit around the moon forever!

There was another frightening note to all this. At one point Lovell was very tired and punched the wrong key on the computer.

This had caused the computer to erase part of its memory and start over. Houston was fearful that a portion of the memory which controlled the reentry might have been destroyed. There were hours of exhaustive checks on the ground to prove that this had not happened. Frank Borman, well aware of the weariness of his crew, scrubbed all further experiments so that everyone would be well rested for TEI.

This preliminary flight caused NASA to make some changes in future flight plans. It was clear that the men needed more periods of rest. NASA had also planned to have the men sleep in shifts so that one astronaut would always be on call in case of a problem. This schedule did not work out. It was decided that all would sleep at the same time, with one man keeping his headset on in case Houston detected trouble on the ground.

On the ninth, and next to the last, orbit of the moon, the astronauts gave another TV presentation. Each offered something of his own impressions of Earth's only natural satellite. Commander Frank Borman spoke first. 'The moon is a different thing to each one of us. My own impression is that it's a vast, lonely, forbidding type of existence, a great expanse of nothing, that looks rather like clouds and clouds of pumice stone.'

Lovell added that the 'vast loneliness of the moon . . . is awe-inspiring. . . . The Earth from here is a grand oasis to the . . . vastness of space.'

William Anders described the lunar sunsets and sunrises. 'The long shadows . . . bring out the relief.' For all three, the voyage had been an epic one.

Then, as the spacecraft crossed the terminator of the moon, William Anders introduced what people back on Earth remembered as the most moving moment of the flight. 'For all the people back on Earth, the crew of Apollo 8 has a message that we would like to send to you.'

'In the beginning God created the Heaven and the Earth. And the Earth was without form, and void; and darkness was upon the face of the deep. And the spirit of God moved upon the face of the waters. And God said, Let there be light: and there was light. And God saw the light, that it was good: and God divided the light from the darkness. . . . And God saw that it was good.'

On 27 December 1968, the crew of the Apollo 8 splashed down in the Pacific after one of the most fantastic and successful voyages in the history of mankind.

Threshold to the Moon

The voyage of Apollo 8 became the threshold for yet a further conquest, the goal of putting a man on the moon. The goal, in fact, had always been a part of the space program. The success of each project put the moon just a little closer. John F Kennedy had announced the goal, and though he did not live to see its achievement, Apollo 8 had convinced NASA that a moon landing was possible. Thus it was, that on 20 July 1969, Apollo 11 achieved one of the greatest victories in the history of exploration.

Years had gone into the preparation for the flight of Apollo 11. And now, here at last came the historic words, 'The Eagle has landed.' The 'Eagle' was the name given to the lunar module which served to carry astronauts Neil Armstrong and Edwin 'Buzz' Aldrin to the surface of the moon. Carried by the command module, *Columbia*, the Eagle looked like nothing so much as a giant metallic insect balanced on slender legs. The landing took place in the Sea of Tranquillity, hence the area around the Eagle was called Tranquillity Base. The CAPCOM back at Houston, who was astronaut Charles M Duke, expressed it for the whole world when he replied to the historic message. 'Roger, Tranquillity. . . . You got a bunch of guys about to turn blue. We're breathing again. Thanks a lot.'

Nearly one quarter of the world's population watched as the grayish figure of Neil Armstrong appeared, clambering down the ladder of the lunar module. As he stepped from the ladder, he remarked, 'That's one small step for a man, one giant leap for mankind.' He moved carefully then, through the moon's black shadows and blinding sunlight.

What is it like, this strange alien world, 240,000 miles (386,160 km) from the Earth? Armstrong described his impressions. 'The surface is fine and powdery. I can pick it up loosely with my toe. It does adhere in fine layers like powdered charcoal to the sole and sides of my boots. I only go in a small fraction of an inch, maybe an eighth of an inch (3 mm), [it had been feared that the men might have to wallow through several feet of dust] but I can see the footprints of my boots and the treads in the fine sandy particles. There seems to be no

Previous spread: Aldrin inspects the footpad of the Lunar Module 'Eagle' on the Sea of Tranquility. The footpad sank into the soil only about two inches, due partly to the weak lunar gravity.

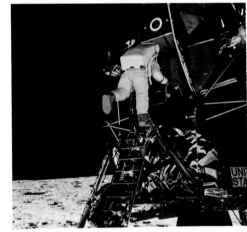

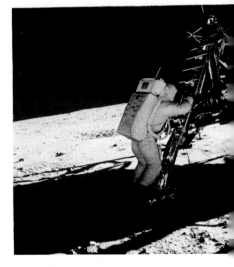

Top: The Lunar Module 3 'Spider' is photographed from the command module.
Above: Edwin Aldrin makes his descent of the Lunar Module ladder.
Left: Astronaut Edwin E Aldrin Jr descends the steps for his moon walk.

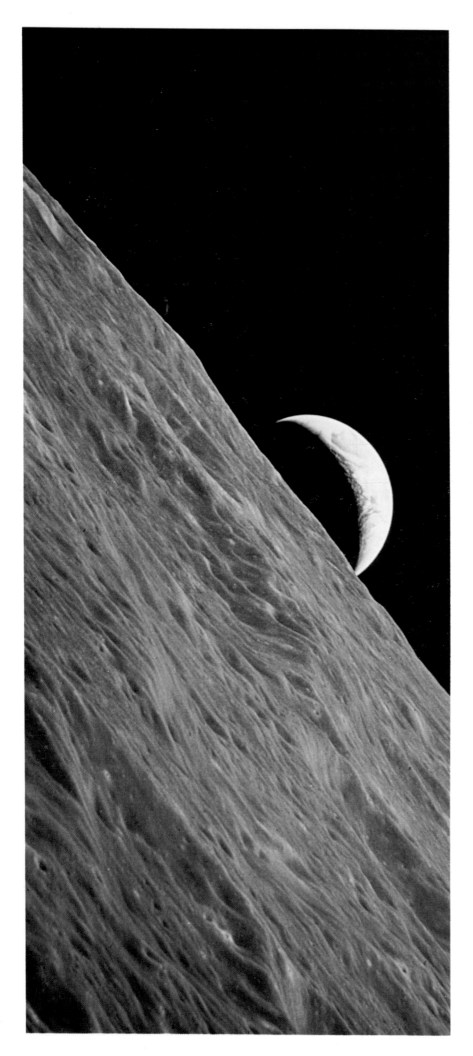

difficulty in moving around. As we suspected, it's even perhaps easier than the simulations at 1/6 g that we performed on the ground.' The 1/6 g of the moon was a real asset to the astronauts. The packs they carried would have weighed five hundred pounds (227 kg) back on Earth. But in the moon's weak gravity, they weighed less than 100 pounds (45 kg). The men found themselves literally bouncing as they walked on the moon's surface.

Nineteen minutes after Neil Armstrong set foot on the moon, he was followed by 'Buzz' Aldrin. Both men remarked on the view. The moon, far from being the forbidding world glimpsed by Apollo 8, 'has a stark beauty all its own. It's like much of the high desert of the United States. It's different but it's very pretty out here.' This was Neil Armstrong's description.

Both men had a bit of trouble finding their center of balance. 'Sometimes,' Aldrin remarked, 'it takes about two or three paces to make sure that you've got your feet underneath you . . .' This was a minor problem, however. Both were able to move about with relative ease, collecting samples of rock and soil and setting up scientific instruments with relatively little trouble.

One of the scientific instruments set up on the moon's surface was a passive seismometer. It contained four seismometers in one. It was powered by solar panels and nuclear heaters in order to withstand the

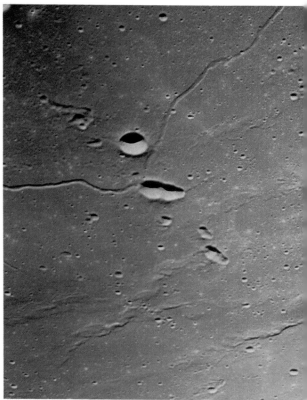

frigid lunar night temperatures. This instrument was so sensitive that it could detect meteorites as small as a pea bombarding the moon from half a mile away. When Armstrong and Aldrin jettisoned their portable life support systems, the seismometer reported the shock as each one struck the moon's surface.

Another instrument set up on the moon was a laser reflector, designed to bounce narrow light beams back to Earth. This would enable physicists to make more precise measurements of Earth-moon distances.

The landing gear of the lunar module would be jettisoned in the return docking with *Columbia*. To commemorate the first moon landing, a small plaque had been attached to it. The plaque read: 'Here men from the planet Earth first set foot upon the moon July 1969, AD. We came in peace for all mankind.' Also left on the moon were a shoulder patch from the Apollo 1 mission to commemorate 'Gus' Grissom, Edward White and Roger Chaffee, two medals to honor the Soviet cosmonauts, Yuri Gagarin and Vladimir Komarov, who had given their lives in their nation's space effort, a silicon disk which had been etched with the goodwill messages from leaders of 73 countries and a golden olive branch which symbolized peace. They were small mementos of a great human effort.

On what was the longest long distance call ever, the astronauts spoke with Presi-

Above: Astronaut Edwin Aldrin inside the Lunar Module after his moon walk.
Opposite left: A crescent Earthrise prior to TEI of Apollo 17.
Below left: An oblique view of the Sea of Tranquility.
Below: Neil A Armstrong inside the LM as it rested on the moon's surface.

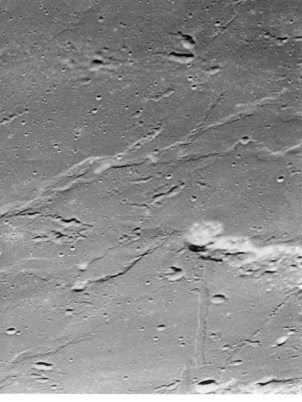

dent Nixon who was in the Oval Room of the White House. The president congratulated them on the success of their mission and added 'For one priceless moment in the whole history of man, all the people of this Earth are truly one. One in their pride in what you have done. And one in our prayers that you return safely to Earth.'

The two men had been on the surface of the moon for nearly two hours, setting up scientific equipment and gathering samples of soil and rock. Fifty pounds (23 kg) of rocks and soil would accompany them back to Earth. To waiting scientists it would be a treasure more precious than gold. But now it was time to leave. The CAPCOM warned that time was running low. Making their way back into the Eagle with a fifty pound (23 kg) load of rocks was a bit awkward but with care both made it through the hatch and were able to repressurize the module. Twelve hours after the Eagle's landing, the television camera on the moon stopped transmitting. Armstrong and Aldrin took time for some much-needed rest. Liftoff from the moon would require them to be totally alert and able to function.

Seventy miles (112 km) above the surface of the moon, Michael Collins orbited in the *Columbia*, watching the activities of his fellow crew members on a television screen. Collins would play a crucial part in the docking exercise that would return the astronauts to Earth. As always, there were a great many concerns over this rendezvous. Would the engine ignite as scheduled? Timing for the docking would be critical.

A minor mishap had caused some apprehension on the part of Ground Control. It had occurred when one of the backpacks barely cleared the hatch as the astronauts re-entered the Eagle. The backpack struck a circuit breaker, snapping off one end of it. That particular circuit breaker was needed to arm the ascent engine. Arming the ascent engine was one of the necessary steps in firing the engine for lunar liftoff.

But, as with nearly every procedure on board the spacecraft, there were at least two ways of accomplishing each maneuver. This was done as a safety precaution. In the case of the circuit breaker, it proved unnecessary since the breaker could still be pushed in. The knowledge, however, that

Above left: The expended Saturn IVB stage as photographed from the Apollo 7 spacecraft during transportation and docking maneuvers at an approximate altitude of 124 nautical miles.
Above : The deployment of the Early Apollo Scientific Experiment Package is photographed by Neil A Armstrong during the Apollo 11 EVA. Astronaut Aldrin is deploying the Passive Seismic Experiments Package.
Opposite top: A closeup of the Apollo 11 Lunar Landing Commemorative Plaque.
Opposite center: The flag of the United States deployed on the surface of the moon dominates this photograph taken from inside the LM.
Opposite bottom: A large boulder with multiple cracks which Astronaut Harrison H Smith described in detail on the Apollo 17 flight.

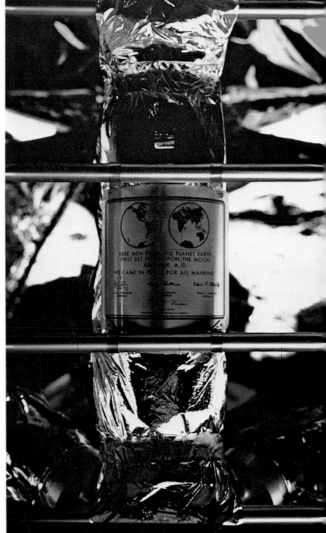

such precautions existed was of great comfort to the men whose lives depended on such tiny devices.

The Eagle rose easily into orbit after less than eight minutes of firing by the ascent engine. Michael Collins, who had kept a lonely vigil from the *Columbia*, watched the proceedings with great joy. From a pinpoint of light the Eagle's tracking beacon grew ever larger, until the historic craft swung into position ready for rendezvous. For a few moments, there was difficulty in getting the two vessels aligned for docking. A little jockeying by the pilots brought the craft together. Michael Collins floated into the tunnel which joined the Eagle and the *Columbia* eager to shake the hands of his jubilant comrades.

Once the three were inside the command module, the ascent stage was set adrift to remain in orbit indefinitely around the moon. Ahead lay a 60-hour journey back to Earth. It was a time of small celebrations. Mike Collins shaved off the beard he had grown in space, but kept his 'moon mustache' as a souvenir of the trip. Aldrin demonstrated for TV audiences the art of

Left: Astronauts Neil A
Armstrong, Michael Collins
and Edwin E Aldrin Jr of
Apollo 11.
Below left: A moon foot-
print photographed on the
Apollo 11 moon walk.
Below right: Astronauts
Tom Stafford and John
Young demonstrate
weightlessness on
Apollo 10.
Bottom left: Astronaut
Edwin E Aldrin Jr and the
US flag on the moon.
Bottom right: Securing the
Apollo 18 spacecraft
following splashdown—
24 July 1975.
Opposite: The liftoff of the
Apollo Soyuz Test Project
Saturn 1B—15 July 1975.

Above: A photograph taken by the second US lunar landing mission of remains of the Surveyor III which had soft-landed on the moon two and one-half years before—photo taken 28 November 1969.
Opposite top left: A view of the Apollo 14 LM, Antares.
Opposite top right: A photo taken during one of the moon walks of the Apollo 12 mission—28 November 1969.
Opposite center: The Lunar Rover used on the Apollo 15 Mission—11 August 1971.
Opposite bottom: A close-up of the Lunar Roving Vehicle at the Taurus-Littrow landing site of Apollo 17.

making an open-faced ham sandwich at zero g. Simply letting the bread float in front of him, he applied ham paste to it and nonchalantly floated it across the cabin to one of his fellow crew-members. The voyage home was as serenely uneventful as the trip out had been. An Air Force pilot who watched the jettisoned service module burning up on re-entry, described it as 'a fiery object . . . with a long tail of orange fire. Out of this hurtling ball of fire came a spectacular shower . . . a mammoth 4th of July display.' In the meantime, the glowing *Columbia* streaked across the skies heading for splash-down in the Pacific, 950 miles (1500 km) southwest of Honolulu. It was not the 4th of July but the 24th. It was not merely an American feat, but a moment of jubilation for all mankind.

The astronauts, emerging from their blackened command module, were at once plunged into quarantine. Wearing coveralls with gas masks, they were transported immediately from their helicopter to a specially adapted vacation trailer known officially as the mobile quarantine facility. They were brought by ship to Hawaii and then by plane to Houston where they were housed in the living quarters of the Lunar Receiving Laboratory. At this point, each man underwent intensive medical tests and examinations. In no case did the men or any of their lunar samples reveal organisms that might be harmful to life on Earth. In fact there were no organisms at all, proof that the moon was indeed a 'dead' planet. After the quarantine period, the astronauts were released to their families on 10 August.

The flight of the Apollo 11 marked the 21st manned flight of the space program. It also marked the 40th anniversary of Robert Goddard's first launched instrumented rocket—a rocket which carried only a barometer, thermometer and a camera. In those mere 40 years, the space age had advanced to an 'impossible' dream. The adventure was the greatest that mankind had ever dared or shared so widely. Most incredible of all was the vast technology that made such an event possible. Behind this single voyage were tons of blueprints, 20 million pages of manuals, 20 thousand contractors and the rocket and spacecraft which consisted of more than five million separate pieces. To consider that all these millions of bits and pieces could be brought together into a successfully functioning whole made the voyage of Apollo 11 all the more miraculous.

Perhaps the moment which best summed up this magnificent accomplishment were when Aldrin read the simple words from the eighth Psalm, 'When I consider thy heavens, the work of thy fingers, the moon and the stars, which thou hast ordained; what is man, that thou art mindful of him?'

There were succeeding flights to the moon, all equally triumphant. But none more certainly captured the enthusiasm of the world than this first 'fantastic voyage.'

Two years later, another moon flight, Apollo 15, would undertake to explore the mountains of the moon. Apollo 15 was an ambitious project. As the 25th manned space flight, it would involve the most extensive exploration of the moon. The crewmembers of the *Endeavor*, as the command module was known, were David R Scott, James B Irwin and Alfred M Worden. The flight had begun on 26 July 1971 with a picture-perfect launch from Cape Kennedy. There had been a few slight problems in the flight, a warning light that showed that a switch had shorted out. The switch controlled the all-important service propulsion engine. There was some concern that it might not fire. But Mission Control at Houston had been ready with solutions. Thus the water leak in the *Endeavor* and the shattered glass cover on a meter in the lunar module, Falcon, were only minor inconveniences.

On the afternoon of 30 July, the Falcon and the *Endeavor* separated. Worden, who would remain alone in the *Endeavor*, took the command module into a higher orbit around the moon. Meanwhile Scott and Irwin, in the Falcon, dropped ever lower to

the moon's forbidding surface. The target for landing was a small basin, rimmed by mountains on three sides and by a deep gorge known as the Hadley Rille on the fourth. The landing promised to be a tricky one. The astronauts would have to come in low over the mountains, brake quickly and make a steep descent to the surface. They would have to stop short to avoid the rille. Despite all these complications, the landing was a perfect one. No matter that the Falcon listed some 11 degrees because one leg was resting in a depression. Nor did the extra ton and a quarter of weight cause a problem. Scott and Irwin were jubilant.

By the time the Falcon had set down on the surface of the moon, it was too late to do much exploring. They were scheduled for a long extravehicular activity (EVA) on the following day. At this point, they would need all the rest they could get. After a supper of cold tomato soup and hamburger, they crawled into their hammocks to sleep.

Sleep aboard this 'portable Leaning Tower of Pisa' was on a par with the supper they had just eaten. The quarters were cramped and the spacecraft made strange noises, made by the fans which circulated oxygen and the glycol pump which forced coolant into the electronic gear. Not the average noises of a household, but Irwin and Scott had practiced sleeping with a tape recording of these noises. They fell asleep just as though these strange sounds were the familiar noises back home.

Nine hours later, fully rested, the men descended to the moon's surface. Radar signals, beamed from the Earth and reflected back by the moon, seemed to indicate that this particular site would be covered with a deep layer of dust. Irwin and Scott found that this was exactly the case. There was nearly a foot (30 cm) of gray powder which clung to everything it touched. Their boots sank an inch or more in the stuff. Irwin said that he was reminded of the soft powder snow at Sun Valley, Idaho, where he had often skied. The dust coated their white space suits and had to be brushed away frequently. The dark gray powder would absorb the solar heat, thus putting an extra strain on the cooling mechanism of the life-support systems.

A unique piece of equipment on this mission was a remarkable vehicle known as the Apollo 15 Rover. It looked like a cross between an electric dune buggy and a

Above: A thin section of moonrock from Apollo 11.
Opposite top: The lunar far side taken from Apollo 11.
Opposite bottom: A view of the Central Station of ALSEP—Apollo 14—
15 February 1971.
Below: This three-frame sequence of the Earthrise over the moon's horizon was
taken from the Apollo 10 Lunar Module, with Astronauts Stafford and Cernan
aboard.

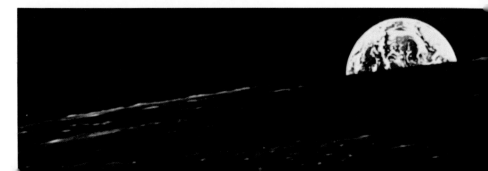

stripped-down jeep. It was, however, far more complex than either of these earthly vehicles. Its tires were made of woven piano wire and faced with titanium chevrons to enable the Rover to ride without sinking into the dust. Each wheel hub contained a sealed electric motor, similar in size to that in a carpenter's electric drill. Two $36\frac{1}{2}$ volt batteries provided power. Added to all this were a gyroscopic navigational system and a radio unit which would allow the astronauts to talk directly to Earth. On Earth the Rover weighed 455 pounds (207 kg), but on the moon its weight was a mere 76 pounds (35 kg). Its sturdy design enabled it to carry two and one-half times its own weight. The Rover had been assembled at the Boeing plant. Much of the work of assembly had been done by hand by skilled engineers. While driving this light vehicle in the moon's 1/6 g took some practice, the astronauts praised it as 'great sport' and 'a

super way to travel.' James Irwin described the ride a bit more graphically as, 'a combination of a bucking bronco and rowboat in a rough sea.' Using the Rover, Irwin and Scott found they could travel at speeds up to seven or eight miles per hour (13 km per hour). Caution was needed, however. At one point they took a downhill turn too fast. As the front wheels dug in, the rear end broke away in a 180 degree skid. Needless to say, riding a moon rover was not exactly your average Sunday drive.

Despite the exuberance of riding the Rover over the moon's surface, the astronauts made a close study of the geology around them. For 14 months before their flight, Scott and Irwin had been carefully trained by Professor Leon Silver from the California Institute of Technology. In a crash course in geology they had learned much about rocks and minerals and the geological history of mountains and volcanoes. They had learned the techniques of observation, reporting and collecting samples. Now, on their own, they gathered samples and information to take back to Earth. The Apennine Mountains which loomed around them resembled nothing they had ever seen on Earth. Unlike their rocky and craggy earthly counterparts, these mountains were smoothly rounded. There were lava flows, craters and meandering gorges. It was this abundance of geological features which had led scientists to recommend the Hadley landing site. Of particular interest was the 650 mile-wide (1046 km) Mare Imbrium, a dark circular area which seems to form the right eye of 'the man in the moon.'

It was hoped by geologists back on Earth that these explorations of the moon would give some hint of the origin of both the moon and the Earth. What, indeed, had caused the Hadley Rille in an airless, waterless environment? Like other Apollo explorers before them, Irwin and Scott would set up a series of experiments and equipment to monitor conditions on the moon. One group of instruments was used to detect tremors and quakes on the surface. Another measured the presence of various gases. Still other instruments checked for

Top left: Astronaut Edwin E Aldrin Jr deploys the Solar Wind Composition Experiment—Apollo 11— 31 July 1969.
Center left: James B Irwin and the Rover—Apollo 15.
Bottom left: The plaque left on the moon by the Apollo 15 crew, listing the names of deceased astronauts and cosmonauts.
Opposite: David R Scott salutes the flag.

the presence of water vapor. There were hundreds of things that scientists back on Earth were curious to know. Four hours on the moon's surface was hardly long enough for the astronauts to set up all the equipment needed. There were thermometers to measure the flow of heat from inside the moon to its surface, a magnetometer to measure the magnetic field, a spectrometer to measure the solar wind (electrified particles constantly blowing at high speed from the sun) and two devices which would measure gases in the moon's thin atmosphere plus a sensor which would measure the accumulating dust. Along with these went a radio transmitter to send the collected data to Earth and a radioisotope thermoelectric generator to provide the power. These would send back messages to Earth for at least five years.

Scientists believed that radioactive elements within the moon would release heat as they disintegrated. The heat would flow toward the surface of the moon. To test for this heat-flow theory, Scott and Irwin would set up a heat probe.

Top left: A view of the full moon photographed from the Apollo 11 spacecraft during its journey home—21 July 1969.
Center left: Astronaut Thomas P Stafford (left) and Russian Cosmonaut Alexi Leonov in the Soyuz Orbital Module during the US-USSR docking.
Bottom left: The Apollo 16 Command and Service Modules as seen from the Lunar Module.
Opposite top: An Apollo 15 ALSEP panorama.
Opposite center: Astronaut Harrison H Schmitt of Apollo 17 working with a lunar scoop.
Opposite bottom: Astronaut Charles M Duke Jr examines the surface of a large boulder.
Below: Astronaut Vance D Brand in the hatchway between the Apollo 18 Command Module and Docking Module—28 July 1975.

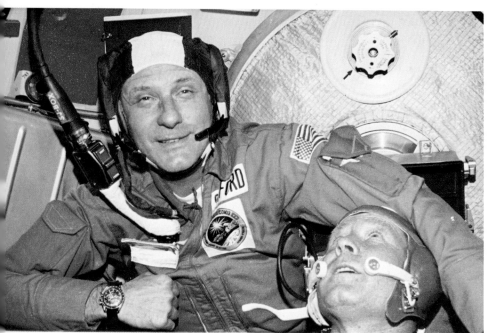

The astronauts found that setting up the heat-flow probes was the most difficult of all the equipment to be installed. Using a small jackhammer drill, David Scott attempted to bore two holes in the moon's crust. The drill, which on earth could bore through three to five inches (8–13 cm) of rock in one minute, slowed noticeably below a foot or two (30–60 cm) of the moon's surface. Soil on the moon was denser than any that the two astronauts had ever encountered. At a depth of five and one half feet (1.7 m), David Scott was forced to stop the drilling operation. The drill would simply go no further. The second hole fared no better.

After six hours of EVA, Scott and Irwin returned to the lunar module. As they removed their suits and helmets, they noticed an odor like gunpowder. They thought that it came from the lunar soil which covered their boots and uniforms. They also noticed that their fingers were sore—perhaps from all the exertion of drilling and setting up the instruments.

Meanwhile, circling above the moon, Al Worden made detailed photographs of the landscape below. Along with the photos, Worden sent back careful observations of the geology seventy miles (112 km) below. He noted evidence of recent volcanism, of gases rising from the moon's interior; all this deduced by a sighting of cinder cones. The cones were mute evidence of volcanic activity—eruptions of ash and gases. They were also further testimony of the primeval geology of the moon. As Worden orbited the moon, he put into action a group of instruments and cameras housed in a compartment of the spacecraft known as the SIM (scientific instrument module) bay. Within SIM were a panoramic camera, a mapping camera, a laser altimeter which would

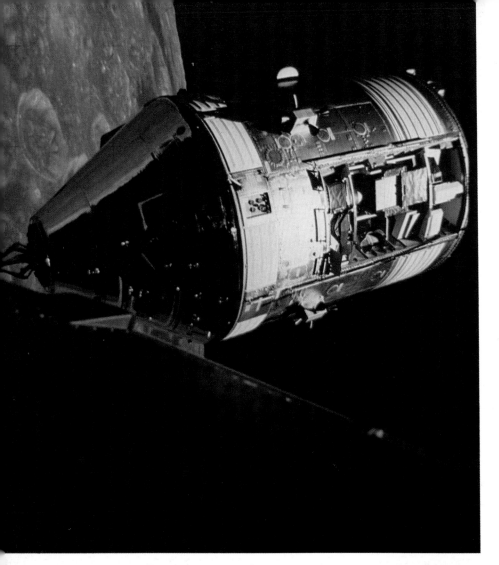

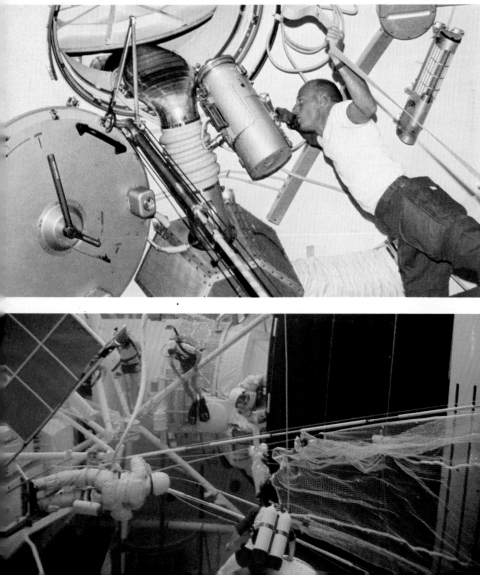

measure surface elevations, a mass spectrometer to measure the thin atmosphere of the moon, and three experiments which would determine the chemical composition of the surface by measuring the emission of gamma rays, X-rays and alpha particles.

One of Al Worden's main problems was exercising amid the narrow confines of the spacecraft. Because prolonged weightlessness causes some difficulties for the bones and circulatory system, Worden had to exercise each day. One of his exercises was running in place. Asked how he accomplished this in the craft, he replied, 'Easy. You just kind of freewheel your legs!'

Seventy miles (113 km) below, Worden's colleagues, Scott and Irwin, continued their exploration of the lunar surface. On this second day of exploration, they were able to collect some milky-white crystalline rock samples which prove to be anorthosite. They were at least four billion years old. Numerous other samples were collected, all in all about 170 pounds (77 kg) of sample rock and soil would be carried back to earth. Perhaps the most difficult feat for Scott and Irwin had been to drill holes in the moon's surface.

By the third day on the moon, time was running short. With great effort, the astronauts were able to pull up a core of moon material measuring more than 90 inches (2.3 m) in length. They gathered specimens of bedrock from the Hadley Rille, and then as a final act placed a memorial to the astronauts and cosmonauts who gave their lives in the effort to explore space.

Then, in the first televised launch from the moon, the small ship lifted off, carrying Scott and Irwin to a rendezvous with crewmate Al Worden and the *Endeavor*. Once back aboard the *Endeavor*, there were yet other experiments to be set up. Set in orbit around the moon was a small 31 inch (80 cm) satellite.

Within the satellite was a gravity experiment to measure the fluctuations of gravity. (Certain areas of the moon seem to exert an extra strong gravitational pull.) Another instrument within the satellite would measure the charged particles within the vicinity of the moon. A third instrument, a magnetometer, would measure the magnetic field of the moon at the level of orbit. As the *Endeavor* moved earthward, the crew took X-ray observations of the galaxy and ultraviolet pictures of the moon and the Earth. Al Worden stepped out in space to recover the film cassettes from the SIM bay.

Though one of the three parachutes collapsed as the command module headed for splashdown, the Apollo 15 crewmen landed safely after a most remarkable and exciting voyage. The treasures they had brought back would furnish scientists with materials to study for many years to come.

Throughout the space program was the often-expressed desire to promote the idea of one world, of world peace and brotherhood. With the final Apollo flight, one small step was taken in this direction. For two years, the crews of the Russian spaceship Soyuz and the Apollo had exchanged visits, learned each other's languages and studied each other's equipment. All of this was done in preparation for a historic rendezvous in space. So it was that on 23 July 1976 Aleksey A Leonov and Thomas P Stafford reached out to exchange a handshake in space. The two ships achieved link-up 140 miles (225 km) above the Earth. As part of the union exercise, the astronauts, Stafford, Vance D Brand and Donald 'Deke' Slayton agreed to communicate only in Russian. The cosmonauts, Leonov (who was the first man to walk in space) and Valeriy Kubasov would converse only in English. Previously, there had been some joking about needing an interpreter for Tom Stafford whose Oklahoma twang was well-known.

With docking latches ready, the Apollo crewmen executed approach maneuvers. With mock nervousness, Leonov joked, 'Please don't forget about your braking engine.' Then over the Atlantic, the two ships joined. Apollo's specialized docking module, located on the front of the ship, allowed the two ships crews to visit each other, but also maintained the different atmospheres of the two crafts. The moment was also an historic one for the Russian people who had never seen live television coverage of a space flight from launch to 'bumpdown.' (Soviet space flights always ended on land, rather than water.)

The link-up was a moment of quiet unity between rival nations. At that moment, perhaps the world recognized the need for cooperation in the demanding challenge of space exploration and the even more difficult challenge of world peace.

Opposite top: A view of the Apollo 17 Command and Service Modules taken from the Lunar Module 'Challenger.'
Opposite center: Charles Conrad Jr transfers equipment from OWS into Airlock Module—Skylab I—27 June 1973.
Opposite bottom: Astronauts training under water.
Right: The deployed US flag.

Houston, We Have a Problem

Apollo 13 was launched on 11 April 1970. It was destined to become an historic flight, though not for the reasons its predecessors were so celebrated. Command Pilot for 'Thirteen' was James A Lovell Jr. His co-pilots were John L Swigert Jr and Fred W Haise Jr, who were both civilians. Apollo 13 looked like another perfect flight to the moon. The liftoff had gone smoothly, without even the slightest problem. At one point in the flight, the ground control complained that the flight was so routine, 'you're putting us to sleep.'

Indeed, there was no sign of trouble whatsoever on that night of 13 April. On a rooftop observatory at the Manned Spacecraft Center in Houston, several engineers were tracking the Apollo spacecraft. One of them had hooked up a telescope to a television set, so that objects in the telescope's field of view would appear on the screen. The men had lost track of the spacecraft and were focusing instead on the giant booster rocket which was trailing Apollo 13 to the moon. The rocket had dumped its fuel and as a result was tumbling end over end. Sunlight glinting off the rocket made it appear to pulse like a variable star. The engineers had just lost sight of the booster when they noticed a bright dot near the middle of the TV screen. Within the next 10 minutes the dot grew to be a dime-sized white disk. The watchers assumed the white disk was a defect in their television set which had been blipping and flickering badly. It was not until the following morning that they realized what they had seen.

At Mission Control Center, dozens of automatic pens were busily scribbling incoming data radioed from Apollo 13. Suddenly, for a space of two seconds, all of them stopped. In the jargon of ground control, such an event is called a 'dropout of data.' It indicates a major problem with either the electrical system or the transmitting system. Incredibly, no one noticed the two second pause.

Inside the Mission Operations Control Room on the third floor of the Control Center, the flight controllers sat in four rows. There was little going on that had demanded their attention. One group, in fact, had been totaling the number of 13s that had occurred in the flight. The launch

Previous spread: A laser reflector left on the moon to enable laser stations on Earth to bounce laser beams from the moon back to Earth and help determine the size, geographic structures, and geodesic features of the moon.

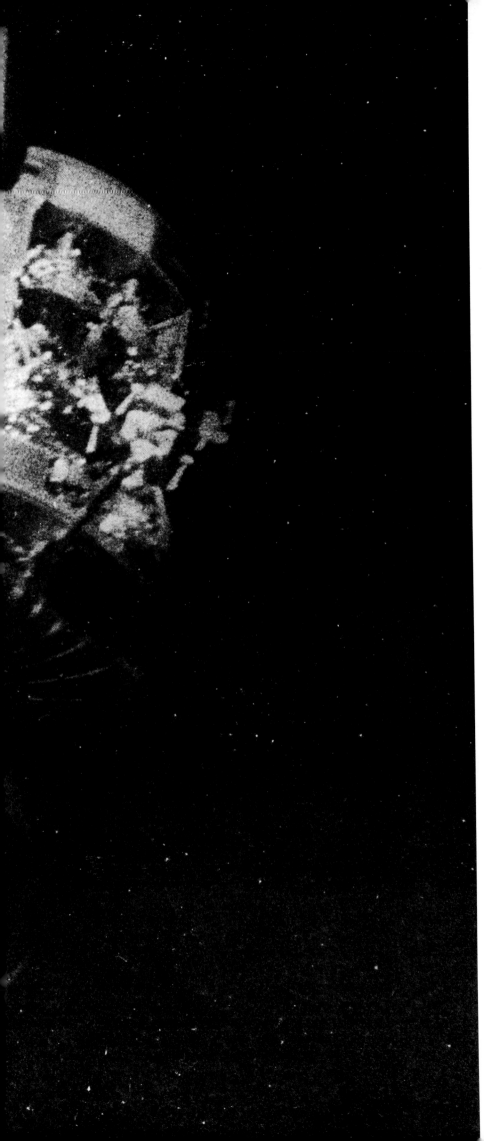

had taken place at 1:13 Central Time, or, in military terms, 13:13.

On board the Apollo 13, the astronauts were broadcasting a television show. James Lovell was acting as cameraman and announcer for the show. He panned the camera around the gray interior of the command module. The module was a cone, measuring about 13 feet (4 m) in diameter and 10½ feet (3.2 m) in height—on the whole as big as a small station wagon. Beneath the command module was the service module which contained the electrical system and the two big oxygen tanks, among other things. Together, the cylindrical service module and the conical command module formed a single, cone-shaped unit. On the front of the cone was the spindly-looking lunar module. In length, the entire unit was almost 60 feet (20 m). Lovell's camera showed the round hatch leading from the command module, Odyssey, to the lunar module, Aquarius. Fred Haise, pilot for the lunar module, would act as guide to Aquarius. He floated through the tunnel into the lunar module, where he demonstrated various kinds of equipment. The televised program concluded with the inspection of the Aquarius. The astronauts re-entered the command module and joined Swigert who had remained at the controls. In front of Lovell and Swigert were two red lights labelled 'Master Alarm.' These would flash on if the computer discovered a serious malfunction. Over their heads was a number of yellow caution lights which would indicate a minor malfunction. At five minutes after nine, one of these began flashing. At the same time a similar light began flashing on the console of the EECOM (one of the Systems Operations Engineers whose duty was to monitor the electrical and environmental systems in the command module). The yellow light warned of low pressure in one of the hydrogen tanks in the service module.

The service module was crammed with equipment. Besides carrying the main propulsion system, there was the system for generating electricity and water. The hydrogen tank was part of this. It was a simple system; the hydrogen and oxygen reacted within the fuel cells, generating electricity. The reaction also generated most of the

Left: This view of the severely damaged Apollo 13 Service Module was photographed from the Lunar Module/Command Module following the SM jettisoning. An entire panel was blown away by the explosion of oxygen tank number two.

spacecraft's water. Like everything else, the system was redundant. There were two hydrogen tanks, two oxygen tanks and three fuel cells. If anything went wrong there was always a backup system.

Since the EECOM, Seymour Liebergot, had been manipulating the hydrogen levels all along, he was not especially concerned by the warning light. But the hydrogen warning did preempt the circuits of the warning system. Thus a problem with the oxygen tanks would not show up in the blinking yellow lights.

Within the service module were 6 compartments that ran the 25 foot (7.62 m) length of the module. The two oxygen tanks with the rest of the electrical generating system were housed in Bay 4. Bay 4 was divided into compartments. Three fuel cells were in the forward compartment. The hydrogen tanks were housed in the rear. In the middle were the two oxygen tanks, silvery spheres made of Inconel, a strong nickel-steel alloy. They were strong enough to contain the oxygen at 900 lbs per square inch (62 kg per square cm). They were

Above: An Apollo 16 view of the Earth—16 April 1972. Although there is much cloud cover, most of the United States and Mexico and some of Central America are visible.
Opposite: A photograph of the Earth taken from NASA's Apollo 13 spacecraft during its journey home. The most visible landmass includes the southwestern United States and northwestern Mexico.

Above: An oblique view of the lunar farside was photographed from the Apollo 13 spacecraft as it passed around the moon on its hazardous journey homeward. This area of the moon is located southeast of Mare Moscoviense.

Opposite: Another view of the damaged Apollo 13 Service Module, with the moon in the distant background. It was photographed from the Lunar Module following SM jettisoning. The Command Module, still docked with the LM, is in the foreground.

made up of an outer and an inner shell. The space between was filled with insulation (some of it flammable). The top of each tank bore a capped dome which sealed an opening for pipes and wires that carried electricity to the fans, heaters and sensors inside the tank.

Had anyone looked inside Tank #2, he would have seen an alarming sight. The wires in the tank were almost bare of insulation. The situation was partially due to an imperfect design and partially to human error. In a countdown test two weeks earlier, the ground crew at Cape Kennedy had filled the tanks with liquid oxygen. Both oxygen and hydrogen were cryogenically cooled to keep them at a low volume. This way they could be stored

compactly. At the time of the test, the oxygen had been at a temperature of −297 degrees Fahrenheit −182 °C). After the test, the ground crew had been unable to remove the oxygen from Tank #2. To force the oxygen out of the tank, the engineers had turned on the heaters and fans inside the tank. As the fans stirred up the oxygen, the heaters would warm it, causing it to expand and be expelled. Or at least, this was their hope. The heaters remained on for eight hours—a period much longer than usual. No one was aware that during that time the heat in the tank was rising higher and higher. There was really no need to worry. The ground crew knew that within the tank was a thermostatic safety switch which would shut off the heat when the tempera-

ture rose to 80 degrees F (27°C)—the safe limit. The thermometer which gave temperature readings for inside the tank did not register above 85 degrees F (29°C)! The major defect here was that the safety switch had been designed to operate on the spacecraft's 28-volt power supply. When the tanks were tested at Cape Kennedy, the power came from a 65-volt supply. It was thought that the liquid oxygen would keep the switch cool. But as Tank #2 emptied, the switch overheated and failed.

This was now the situation as Fred Swigert prepared to stir the tanks, He had had trouble all day getting an accurate reading on Tank #2. Now he hoped to be able to get some readings on the pressure, temperature and quantity of oxygen within

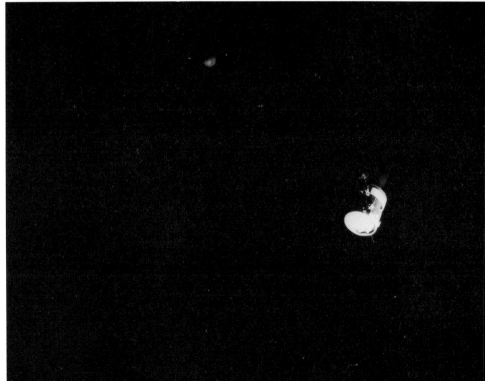

the tank. For sixteen seconds all was quiet. Suddenly in tank #2 an arc of electricity shot between two bare wires. Within seconds, the pressure in the tank began to rise as the arc heated the oxygen. Because the astronauts and the ground control were focused on the low-pressure reading in the hydrogen tank, no one realized what was happening in Tank #2. The yellow warning lights for the hydrogen tank preempted every other light in the system.

At eight minutes past 9 pm, the pressure of the expanded oxygen blew the dome off the tank. The insulation between the inner and outer shells of the tank caught fire. The heat in Bay 4 was like a blowtorch. The silvery Mylar, which lined the inside of the bay, ignited. Gases, building up within the bay, blew off its cover. It was sheer luck that the cover blew off. Had the pressure been allowed to build, the command module, riding as it did on the front of the service module, might have been blown off like a cork from a bottle. The three men aboard the spacecraft were aware that something had happened. But in the vacuum of space it was difficult to describe exactly what had taken place. Without air to carry sound or shock waves, such an explosion was less noticeable than it would have been on Earth. There had been a shudder, a bang, a kind of vibration which went through the spacecraft. John Swigert, strapped in his seat, had felt a distinct shudder. James Lovell, floating above his seat, had heard a bang, which he thought was Haise opening a valve in the lunar module. Fred Haise, coming out of the tunnel between the command module and the LM, noted with apprehension the jolting movement. Normally the tunnel shook from side to side. This time the movement was from end to end. John Swigert's voice on the radio was so calm that for a moment the CAPCOM, Jack Lousma, did not recognize who was speaking. 'Houston, we have a problem.' Only the Flight Surgeon realized the calm was artificial. The pulse reading for all three astronauts had shot up from 70 to over 130 in those first anxious moments.

No one, at least no one on the ground, could believe that anything more than an instrumentation failure had occurred.

Yet within a minute after the explosion, Lovell was reporting to the ground crew that the Main Bus B had no power in it and power in Main Bus A was dropping as well. The EECOM, Liebergot, was thoroughly confused. The two main buses drew their power from three redundant fuel cells. If one lost power, the other should be holding up. Clearly there was a problem with the spacecraft's electrical system, but where did it lie?

The lights on the telemetry screen in front of Liebergot blinked; half of them amber. Such an incident had happened only once before when Apollo 12 had been struck by lightning during liftoff. This was certainly not the cause now.

What had happened to Main Bus B? He stared at the diagram of the electrical system on the console in front of him. There were the hydrogen and oxygen tanks linked to the fuel cells which were in turn linked to the buses. Linked to the buses was all the equipment within the spacecraft that was supposed to be receiving power.

Fuel Cells 1 and 2 were supplying Main Bus A which was still functioning. Main Bus B drew all its power from Fuel Cell 3 which had stopped generating power. With consternation, Liebergot learned that Fuel Cell 1 was no longer functioning either. The spacecraft had lost two fuel cells. It was unheard of. The power in the remaining good fuel cell was beginning to drop, too, though not yet enough to be of concern.

The two dead fuel cells drew their oxygen from the same two tanks as did the one good cell. Therefore, Liebergot did not give the oxygen tanks much thought. The tanks, though not exactly redundant, might as well have been. There were so many safety valves between them. There were also valves between the tanks and the fuel cells called reactant valves. These could be shut off from the oxygen, too. The system of valves was especially critical. Along with fueling the electrical system and producing water, the oxygen tanks provided the command module with oxygen for breathing.

It was like some horrible grade B movie. The nation's first accident in space had occurred. But no one knew just what had happened. Two dead fuel cells! It was impossible. NASA planned every system with a backup system. Backups didn't fail. There had to be a simple explanation somewhere. Perhaps, Liebergot speculated, the jolt or bang or whatever it was had knocked the fuel cells loose from the buses. He suggested the astronauts run a check to see if the cells were still hooked up. There was no change in the electricity level of the buses.

Right: An interior view of the Apollo 13 Lunar Module showing the 'mail box' that eliminated CO_2.

The trouble was nothing so simple as broken connections.

Within the spacecraft were two identical balls set in the dashboard. They were the flight-director attitudes. Like a three-dimensional compass, they showed which way the spacecraft was heading. When the jolting bang occurred, they appeared to spin in an odd, erratic movement. Actually the movement was caused by the pitching and yawing of the spacecraft. The guidance computer which automatically fired the 16 small thruster rockets to hold the craft steady, was unable to stop the pitching. Lovell was attempting to control the wobble by firing the rockets manually. He was having no more success than the computer. The spacecraft continued to wave and pitch as if something were spewing out from it, giving it a strange kind of thrust.

Despite all the suggestions from Ground Control, Lovell was unable to control the spacecraft's attitude. Guidance and Navigation Control was especially anxious that the spacecraft be steadied. Once Lovell was able to control it, he could set up the easy, passive thermal roll. Under this system, the spacecraft would roll once every twenty minutes. In this way the temperature could be controlled from extremes of heat and cold. The roll enabled uniform temperatures on all sides. Some of the spacecraft's propulsion systems were very sensitive to temperature extremes.

Yet another problem caused by the wobbling was the fact that if the spacecraft rolled into certain attitudes the guidance system would lock. A spherical structure called the inertial measurement unit was the heart of the guidance system. At the foot of the center couch in the lower equipment bay was the guidance platform. Before the launch, the platform had been aligned with certain stars. It swung freely on three gimbals enabling a set of gyroscopes to maintain its position relative to the stars. No matter how the spacecraft pitched and yawed, the system of gimbals and gyroscopes would keep the platform aligned. The danger was that if the gimbals lined up in a certain way, they would lock. The spacecraft would then be without any reference point in space—an event equivalent to being without a compass! Time after time, the spacecraft wobbled toward gimbal lock. Each time, Lovell would quickly change the direction of the craft.

The wobbling flight was also affecting radio communications. The antennas had to be directed at the Earth to receive signals. Even with the four big omnidirectional antennas there were moments when communications stopped completely.

The astronauts were so busy maintaining their guidance system, transferring the thrusters and checking the antennas, that they had little time to worry about what trouble was facing them. Back at Mission Control, however, the TELMU or LM systems officer, Robert Heselmeyer, had time to think about the situation. As he sat at his console he watched the little electricity which warmed instruments on the lunar module drop. When it finally stopped altogether, he reported it to Flight Director (FIDO) Eugene Kranz. Kranz already had his hands full and asked Haselmeyer to get back to him later. Concern continued to gnaw at the TELMU. If anything was seriously wrong with the command module, then the lunar module might have to become a lifeboat. While he considered this possibility, he searched his console for instructions on such a procedure.

In an effort to pinpoint the trouble, Liebergot and Kranz had asked Lovell to give the readings on all the gauges which dealt with the electrical system. Lovell had gotten as far as the pressure gauges on the oxygen tanks. Oxygen Tank #2 was reading 0. Anxiously, Lovell floated out of his couch and pressed his face to the window looking backward into the service module. He saw a thin cloud of white vapor. 'It looks to me we are venting something,' he told CAP-COM. 'We are venting something out into space.' This explained why they were unable to control the spacecraft. The venting created the same sort of thrust that a rocket would. Jack Lousma listened to the report in horror. The substance venting was obviously a gas (a liquid would have formed hard nuggets) and the gas must be oxygen. Tank #2 was at 0 and the pressure in Tank #1 was dropping. What Lovell was seeing was the oxygen from #1.

Most of the flight controllers still refused to believe what was happening. If there was a leak in the oxygen tank it was probably a slow one, Liebergot thought. What none of the crew on the ground realized was that there had been an explosion. Tank #2 had been completely ruptured. In the explosion, pipes and valves between the two tanks had been ripped out and the resulting break was now leaking oxygen from Tank #1. The supply of electricity for the command and service module would last only as long as

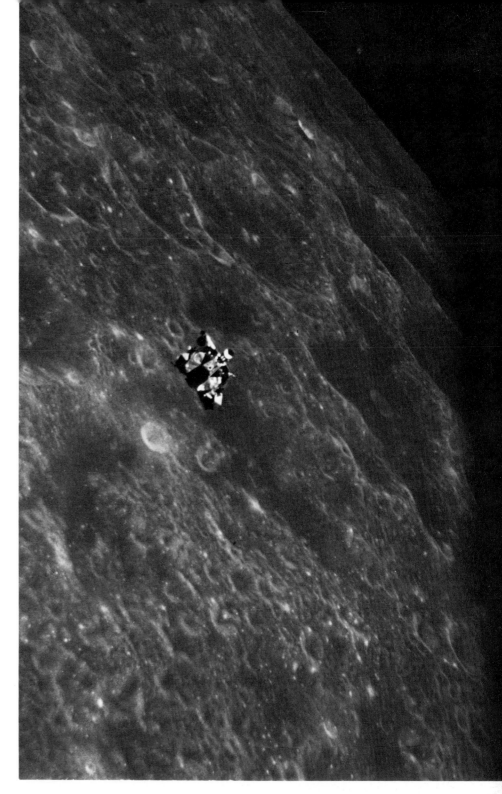

that oxygen in Tank #1 could reach Fuel Cell 2.

While Mission Control was still being cautious in its assessment of the problem, the astronauts realized exactly what was happening. It was important if at all possible to save the moon-landing mission. Haise, Swigert and Lovell knew already that the moon landing was out of the question. Liebergot and Kranz were intent on saving it if possible. To put less strain on the good bus, they decided to have the crew power down the command module. They requested that the astronauts work through the first half of page 5 of the Emergency

Above: The Apollo 16 Lunar Module ascent stage approaches the Command and Service Modules above the rough terrain on the lunar near-side. Several craters in the MacLaurin Complex are near the center of the photo. The crew of Apollo 16 consisted of John W Young, commander; Charles M Duke Jr, Lunar Module pilot and Thomas K Mattingly II, who took the photograph.
Opposite: The Apollo 13 liftoff—2:13 pm (EST), 11 April 1970.

Power-Down Checklist. This checklist explained how to turn off the equipment in proper sequence. It would certainly ease the power crisis. But the astronauts, in order to find the proper checklist, had to search through 20 pounds (9.6 kg) of instruction materials. Despite the efforts at conservation, power in the command module continued to drop. Seymour Liebergot was beginning to see just how serious the problem was. It would not be long until the command module lost all power. When that happened, the astronauts would have to rely on the LM to bring them back to Earth.

Could the LM, designed to support two men for two days, support three men for four days? There was the further problem: the LM was not equipped with a heat shield. Somehow, before the astronauts entered the Earth's atmosphere, they would have to figure out a way to fly the powerless command module. Never in all the pre-flight simulations had anyone imagined such a situation. Eugene Kranz called for extra tracking equipment to be added to the global network which normally monitored each spaceflight. Additional computers at the Goddard Space Flight Center in Maryland were hooked up. On the ground floor of the Operations Wing, the Real Time Computer Complex was asked to bring in one of the big IBM computers which would process data from the spacecraft in 'real time,' that is, instantaneously.

The Retrofire Officer (RETRO), Bobby Spencer, who would be in charge of the emergency return plans, felt strongly that the spacecraft would have to go around the moon to return home. The spacecraft was now 45,000 miles (72,000 km) from the moon and five times that distance from the Earth. It was impossible now to turn back and make a direct return to the Earth. Heselmeyer, the TELMU, had opted for a direct abort, for he felt that there was little chance the LM consumables would last four days. The RETROs were not happy with the idea. Charles Dieterich, who had now joined Spencer, told Heselmeyer that in order to blast out of the moon's gravitational pull, the LM would have to be jettisoned.

For a time, however, the men at Mission Control considered the idea of a direct abort. It was finally scrapped because no one knew anything about the condition of the main rocket in the service module. The spacecraft would take the path around the moon. There was some relief at this decision. It would now be possible to concentrate all efforts on using the LM.

Back on the spacecraft, Fred Haise moved through the tunnel to the lunar module. The LM was smaller than the command module. The dashboard panels, smaller than those of the command module, had many of the same instruments. There were no seats. Each of the astronauts would have to stand.

How a third astronaut would fit into the limited space was difficult to imagine. The problem Haise was facing at that moment was how to power up the LM. The three checklists he found were all based on utilizing power from the command module. Finally the TELMU came up with a set of instructions for powering up the LM on its own batteries. Power in the command module was dropping rapidly. It was important to get the LM functioning very soon. There was no time for the long, complex checklists now. Haise and the TELMU improvised, moving back and forth between various checklists. Luckily, they found they could operate very smoothly. It was a skill they were going to have to rely on for the next few days. Lovell now came aboard the LM.

Below, alone in the command module, Swigert began turning off most of the thrusters and the pumps for the fuel cells. He would leave on the cabin lights, the radio, the guidance system and the heaters and fans inside Oxygen Tank #1. The heaters in the guidance system must also be left on, for if it became too cold the system might not work when it was needed. In four days, it would have to guide the command module through the atmosphere. Power for the heaters would have to come from the LM.

There was still one more function that must be performed before operations could be transferred entirely to the LM. The guidance computer aboard the command module would have to transfer the platform alignment to that of the LM. Once the computer aboard the LM was set up, the astronauts knew they had taken a major step toward getting back home. Three hours after the explosion, Swigert made his way from the dead command module into the LM.

The spacecraft now swung into a trajectory that would bring it around the moon. Rounding the moon would take 20 hours and would be one of the critical points in the flight back toward Earth. During the 20 hours, the flight controllers would have to come up with a plan for reentry and splashdown. The present trajectory would carry the Apollo 13 around the moon and back toward the Earth. But the venting oxygen had caused the spacecraft to leave its free return trajectory. If it held its present course, it would miss the Earth by 40,000 miles (64,360 km)!

Left: The Apollo 9 LM viewed from the CPM.
Opposite: The full moon as viewed from the Command and Service Modules of Apollo 16.

Above: Dr Donald K Slayton (center foreground) talks with Dr Wernher von Braun (right) at an Apollo 13 post-flight debriefing post-flight debriefing session. The three crewmen of the problem-plagued Apollo 13 mission in the background (left to right) are James A Lovell Jr, commander; John L Swigert Jr, Command Module Pilot and Fred W Haise Jr, Lunar Module pilot.
Opposite: Dr Robert R Gilruth (standing, far left), Director of the Manned Spacecraft Center at Houston, is introduced by Dr Thomas O Paine, the NASA administrator, during special Apollo 13 post-mission ceremonies with President and Mrs Richard M Nixon—18 April 1970.

The problem, then was how to correct the course of the spacecraft. Several suggestions were offered. It was finally decided to make a small burn within the next hour or two. This would bring the spacecraft back to the free return trajectory. Eighteen hours later, when they had rounded the moon, the astronauts would conduct another burn to bring the craft into a more precise landing trajectory.

There were more problems in getting everything lined up. No one was sure that the main rocket on the LM would operate. Finally at 2:43 am, James Lovell pushed the button for the rocket burn. For 30 seconds, the astronauts felt themselves being pushed toward the floor. It was the only indication they had that the rocket was firing... another step on the long way home.

At Mission Control the TELMUs were concerning themselves with the consumables aboard the LM. Consumables were water, oxygen and electricity. Although they hoped the trip home would take considerably less than the projected four days, they had to plan for the worst. It was necessary, therefore, to conserve as much as possible. Once the rocket burn was completed, the TELMUs requested the astronauts to power down equipment, especially the guidance computer whose gyroscopes used a great deal of electricity. This prompted arguments from the other flight controllers who were fearful of losing the platform alignment. It was finally agreed that the guidance computer would continue to operate. This made rationing water more difficult, since water was used for cooling the electronic gear.

An even worse problem which the TELMUs faced was the buildup of carbon dioxide within the LM. In normal circumstances, the air in the command module and the LM was constantly circulated through

pellets of lithium hydroxide. (Lithium hydroxide has an affinity for carbon dioxide.) The lithium hydroxide was contained in canisters which fitted into the ventilating system. Once the canister was saturated with carbon dioxide, it had to be replaced. There weren't enough canisters in the LM to last the voyage home. Though there were additional canisters in the command module, they were of a different size and wouldn't fit the LM's system. The TELMUs turned the problem over to the Crew Systems Engineers who were in charge of such equipment.

As word of the disaster got around, many astronauts came into the Control Center to volunteer their help. Many took over the simulators to test maneuvers never tried before. Up in the Apollo 13, Lovell and Haise were attempting to establish a thermal roll. It was a little like having the tail wag the dog. The mass of the service module and the command module was more than twice that of the LM. It had never been meant to control the attitude of the entire spacecraft. On the ground, Charles Duke experimented with a simulator to see if firing the thrusters in short bursts might achieve the roll. On the simulator, at least, it seemed to be effective. Lovell and Haise were fighting fatigue. Finally, Lovell was able to get the spacecraft stabilized and pointed in the right direction. He would have to rotate the spacecraft 90 degrees every hour by hand. But this would achieve the needed thermal roll. Thanks to Duke in the simulator, a couple of big problems had been solved.

It was now up to the flight controllers to make some plans for reentry and splashdown. In its present trajectory, Apollo 13 would splash down in the Indian Ocean, in an area never before used for a landing. The nearest rescue ship was a destroyer cruising off Mauritius 700 miles (1125 km) from the rescue site. It would have to return to port to be fitted with the special crane used to recover command modules. Timing would be a problem.

Another possibility was a fast burn which would bring the spacecraft down in the Pacific, an area where there were many recovery vessels. The one drawback was that the service module would have to be jettisoned before Apollo 13 reached the Earth's atmosphere. It would be impossible for the LM to reach the required velocity otherwise. It would seem a simple enough matter to jettison a useless piece of equipment. But the service module fitted over the

heat shield, insulating it from the extreme cold of outer space. No one knew what effect exposure to that cold might have on the ceramic shield.

There was also the possibility of doing a slower burn which would cost an extra day. By doing the slower burn, the service module could be retained, with the option of jettisoning it later should trouble arise. In the end, the slower burn was the choice.

Then there were the further problems of reentry. On reentry, the astronauts would be back in the command module. At that point they would have to jettison both the service module and the LM. Since the command module was dead, the LM would have to perform all the tasks including jettisoning itself.

The spacecraft had now rounded the moon and was heading back toward Earth. It was traveling at a speed of 5000 miles per hour (8000 km per hour). The time for the burn was approaching. Lovell made ready. On the count of 40 seconds he turned on the main rocket.

Once the burn was conducted, the CAPCOM expected to read up the procedures for powering the spacecraft down. The astronauts had gone nearly two days without any sleep. Brand, the CAPCOM, was anxious to complete the powerdown and get the men some rest. Instead, a debate ensued between the TELMUs, anxious to powerdown, and the design engineers who insisted that a thermal roll must be established. The engineers feared some part of the craft would overheat in the sunlight and break down. In the end, the engineers won out. But establishing thermal roll was immensely frustrating. Time after time the spaceship wobbled out of its roll. To make matters worse, there was an error in the trajectory which was growing. It seemed that once again something in the spacecraft must be venting.

The Crew Systems Engineers had finally come up with a method to clean the carbon dioxide out of the LM's atmosphere. It was not a moment too soon, for a yellow caution light was indicating a dangerous build-up of carbon dioxide.

Left: The Apollo 13 crew—James A Lovell Jr John L Swigert Jr and Fred W Haise Jr.

The spacecraft trajectory continued shallow. It was clear that something would have to be done to correct it or Apollo 13 would miss the Earth entirely. Accordingly a mid-course ban was planned. Shortly before 10 pm on Wednesday, 15 April the crew of the Apollo 13 began preparing for the burn which would correct the attitude of their flight. With all the automatic equipment powered down, everything would be done manually. At 10:31 the burn was complete.

Conditions aboard the spacecraft were rapidly growing more uncomfortable. The temperature in the command module was 38 degrees F (3.3 °C). Haise, ill with a kidney infection, could not control his shivering. Swigert had spilled water in his shoes two days earlier and still had wet feet. The walls and windows dripped with water which was condensing from the cabin air. The cold made it impossible to sleep. And the astronauts were not drinking enough water. They could survive the cold and exhaustion but the lack of water was dangerous. Dehydration could alter body chemistry in such a way as to impair thought and movement.

It was now Thursday. Splashdown was set for noon on Friday. Through herculean efforts, Mission Control had come up with a checklist to use for reentry. Such lists normally took about three months of preparation. The team at Mission Control had done it in three days!

Then it was Friday. The astronauts were only 58,000 miles (93,000 km) from Earth. One small correctional burn was almost too much for the weary men. The LM was now supplying power to the command module. The tension at Mission Control was building. Then it was time for reentry. Silence at Mission Control as everyone waited to hear the astronauts report in. Fully a minute went by. Then came John Swigert's voice, 'OK, Joe.' It was 12:07 pm. Nearby was the *Iwo Jima* waiting to pick up three very weary, very happy men. The Odyssey was home!

An overall view of the crowded Mission Operations Control Room in the Mission Control Center at the Manned Spacecraft Center during post-recovery ceremonies for Apollo 13 aboard the USS *Iwo Jima* —17 April 1970.

Beyond the Moon

Exploration of the moon was but one phase of NASA's space program. Equally ambitious was a series of unmanned spacecraft designed to explore the planets which orbit the sun. Among the earliest was the Mariner program, in which a series of instrument-laden craft were launched to fly by, photograph, record and transmit as much data back to Earth as the equipment would allow. Mariner 9 was the first to achieve an orbit of the planet Mars.

Mariner 9 was launched on 30 May 1971 at a time when Earth and Mars were in the closest proximity to each other since 1924. Carrying barely 150 pounds (68 kg) of scientific instruments and looking like a gone-astray windmill, the blue-winged satellite was spectacularly successful. During its first year of operation, Mariner 9 sent back over 50 billion bits of information via radio signals. These signals were picked up by a huge radio telescope at Goldstone in the Mojave Desert of California. From Goldstone, the signals were sent to the Jet Propulsion Laboratory in Pasadena, California. Here, they were reconstructed as vital information on Mars. Mariner 9 was able to photomap the entire surface of the planet, probe its atmosphere, record its temperature, and determine its chemical make-up. During that first year, Mariner sent back signals which were translated into more than 7300 photographs. Some of these maintained a resolution so fine that objects as long as a football field could be detected. Never before had Mankind had such a close view of the 'Red Planet.'

Earlier fly-bys, in 1965 and 1969, had suggested that Mars was a 'dead' planet much like the moon with a monotonous terrain. There were no mountain ranges, no volcanic activity, no great geologic faults. All of these earlier conclusions were disproved by the information sent back by Mariner 9. The satellite circled Mars twice each Earth day. Using two cameras, one equipped with a wide-angle lens and one with a narrow angle, Mariner 9 photomapped the entire planet. The photographs turned out to be amazingly detailed, showing huge volcanoes larger than any on Earth, great canyons that extended for thousands of miles, dusty basins spreading in all directions and extensive lava flows.

Right: The TV team in the Video Display Room receiving photo prints from the Mariner 9 Mars Mission.
Previous spread: An artist's conception of Venus' largest inland region with an outline of the US for scale.

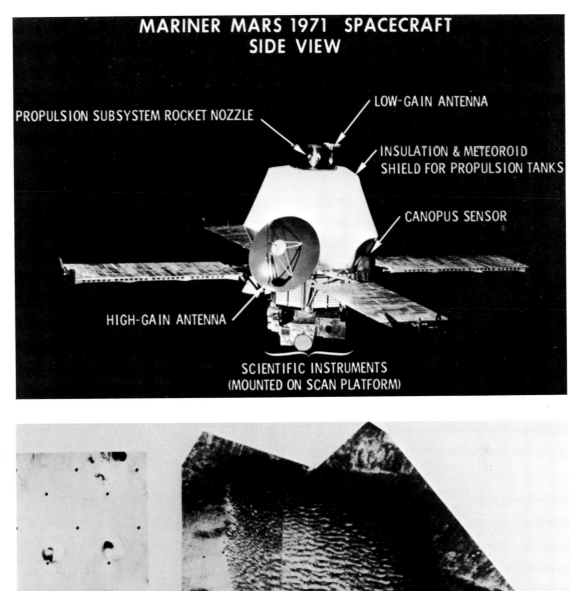

MARINER MARS 1971 SPACECRAFT
SIDE VIEW

PROPULSION SUBSYSTEM ROCKET NOZZLE

LOW-GAIN ANTENNA

INSULATION & METEOROID
SHIELD FOR PROPULSION TANKS

CANOPUS SENSOR

HIGH-GAIN ANTENNA

SCIENTIFIC INSTRUMENTS
(MOUNTED ON SCAN PLATFORM)

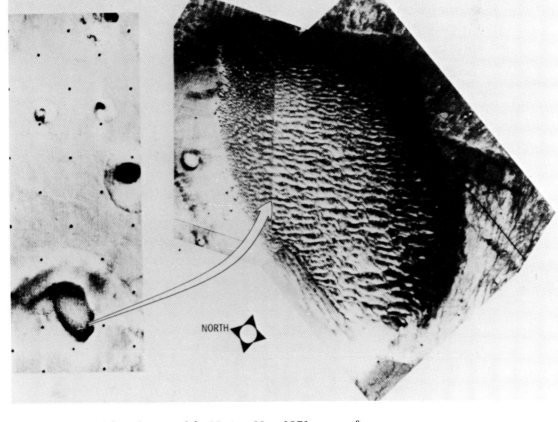

NORTH

Top: A diagram of the side view of the Mariner Mars 1971 spacecraft.
Above: A mosaic of Mariner 9 pictures of Mars (right) shows the wave-like surface texture of the photo left.

Surface elevations on Mars were revealed by an instrument called a spectrometer. As Mariner 9 scanned a track of rugged Martian landscape, the spectrometer would detect changes in the pressure of the thin atmosphere. These changes in pressure could be translated into land elevations.

Still another spectrometer was able to measure atmospheric and surface temperatures. Yet another spectrometer called an infrared interferometer (IRIS) measured the thermal energy radiated by Mars. It could also indicate the intensity of radiation in various parts of the infrared spectrum. This

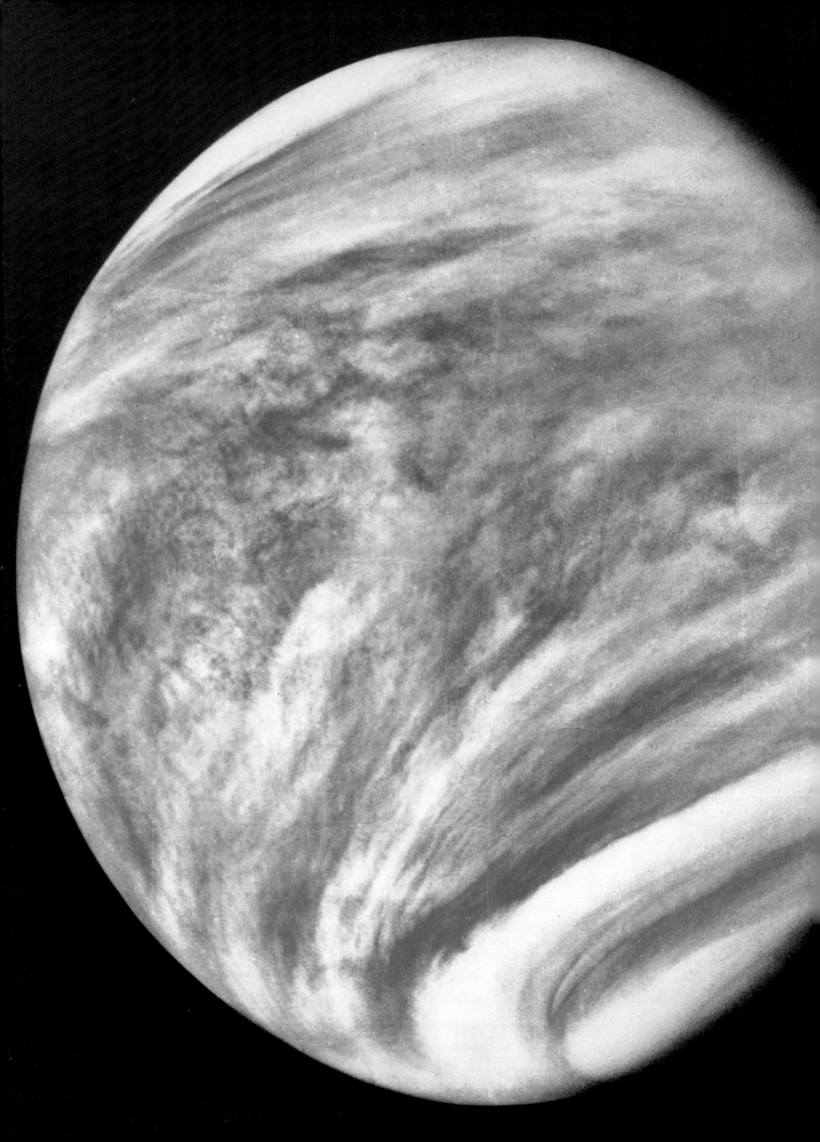

was a further aid to identifying the chemical substances which exist on Mars. The dips and peaks on an infrared reading offer clues as reliable as fingerprints to chemical substances. Because of the great variability in temperature, scientists found that Mars was subject to fierce shrieking winds, perhaps rising to as much as 300 miles per hour (483 km per hour).

These small unmanned spacecraft offered much information to science. Mariner 10 gave the first close-up view of Venus and Mercury, two planets which had been especially enigmatic. The little spacecraft, which looked like a louvered pillbox equipped with radiating arms, solar panels and a 'parasol' sunshade, had had a number of problems. But despite all, it had reached its target with an error of only 104 miles (167 km) in a flight path of 250 million miles (402,000,000 km). Subjected to heat five times more intense than any on Earth, yet freezing within its own shadow, Mariner 10, nevertheless, performed without a flaw.

Mercury, as the planet nearest the sun and hardly larger than the Earth's moon, had always been especially difficult to view. Venus was almost as obscure, due to the heavy cloud cover. For years, astronomers had thought of Venus as the Earth's twin since the two planets resemble each other so closely in size, mass and density. Here the similarity ends. Scientists found no water and no free oxygen exist on Venus. The atmosphere is a thick, heavy layer of carbon dioxide with a pressure equal to the ocean at a 3300 foot (1006 m) depth. The surface temperature is about 750 degrees Kelvin or 900 degrees Fahrenheit.

Mariner 10 showed that the clouds which surround Venus extend far above the surface of the planet. While Earth's clouds usually rise no higher than ten miles (16 km) above the surface, those of Venus rise nearly 40 miles (65 km) high, with a thin haze rising perhaps another 15 miles (24 km). The tops of these clouds seem to be composed of sulphuric-acid droplets!

Detailed ultra-violet pictures of the clouds of Venus showed that, at the equator, they move at speeds of two hundred miles (320 km) per hour. This compared with the slow rotation of the planet (once every 243 Earth days) created a challenging mystery for astronomers.

The gravity of Venus pulled Mariner 10 in a pathway toward Mercury. Mercury, perhaps the smallest of the planets and closest to the sun, was also the least known of all the nine planets. Mariner 10 revealed it as a 'two-faced' planet much like the moon and Mars. The one side of Mercury presented a heavily-cratered surface resembling the geography of the moon. The far side was less heavily cratered and showed areas of volcanic flooding. No atmosphere was detected, but a fairly strong magnetic field seemed to exist. This was something scientists had not expected to find. In this feature, at least, Mercury resembles the Earth. But it is the only similarity. Mercury, at some points in its orbit, is only 29 million miles (47,000,000 km) from the sun. Thus the surface temperatures of the planet vary from 800 degrees Fahrenheit (427°C) at mid-day to −300 degrees F (−184°C) on the dark side. Thus the temperature range on this tiny cinder of a planet is 1100 degrees Fahrenheit (611°C).

One curious feature which Mercury showed was a series of huge curving cliffs. Neither the moon nor Mars showed anything similar to these 'rupes,' as scientists have called them. It is known that Mercury has a very dense core, almost as dense as the Earth's, although its mass is only 6 percent of that of the Earth. Scientists believe the strange cliffs were formed when the dense core shrank slightly early in the planet's history.

The Mariner program began in 1962, with two Venus probes. Mariner 1 had veered off course during launch and had to be destroyed. Mariner 2 successfully completed its mission. The spacecraft were launched at Cape Kennedy, using an Atlas/Agena rocket as the liftoff vehicle for the spacecraft. The payload for the first of these spacecraft consisted of numerous instruments for measuring particles and magnetic fields, plus a single television camera which would photograph the surface of the encountered planet. There were also occultation experiments which made use of an on-board radio. Mariner 3's protective shroud failed to separate from the craft after launch, making that a failure. Mariner 4, launched 23 days later in November 1964, was successfully launched on a course for Mars.

Left: This view of Venus was taken from 450,000 miles away by Mariner 10's television camera 6 February 1974. The predominant swirl at the South Pole can be seen. Individual TV frames were computer-enhanced, then mosaicked and retouched. They were taken in ultraviolet light.

Shortly after midnight on 15 July 1965, Mariner 4 took the first close-up pictures of the planet Mars. Twenty more followed as Mariner flew by Mars. Then the craft passed behind the planet. As it did so, radio signals passed through the Martian ionosphere and atmosphere. This occultation experiment would allow scientists to study atmospheric pressures, densities and temperatures on Mars. Because Mars lies at such a distance from Earth, 186 million miles (more than 300,000,000 km) the data came back very slowly and were not completed until nine and one-half days later. But the data were more revealing than anyone had thought. What was noticeable at once were the lunar-like craters which most scientists had not expected. The occultation experiment showed a surface pressure of from 5–10 millibars, much lower than expected. The atmosphere appeared to be composed mainly of carbon dioxide. As is often the case, Mariner 4's data aroused more questions than it answered.

NASA authorized a second and more ambitious Mariner flight to Mars. For this second flight, two spacecraft, bearing two-camera imaging systems and more complex instruments, were launched. Although the launching dates for the two spacecraft were more than a month apart (24 February and 27 March 1969), their trajectories were planned so that they would encounter Mars just 5 days apart. These were designated Mariner 6 and 7.

On board each craft was a camera with a wide-angle lens and a camera with a telephoto lens. In addition, each Mariner carried an ultra-violet spectrometer, an infrared spectrometer and an infrared radiometer. The spectrometers would detect and

Right: Signals from the Mariner 9 spacecraft were fed into this high-speed digital computer for storage and transmission.
Below: The Mariner Mission Control Center activity at the time of orbit insertion—30 May 1971.

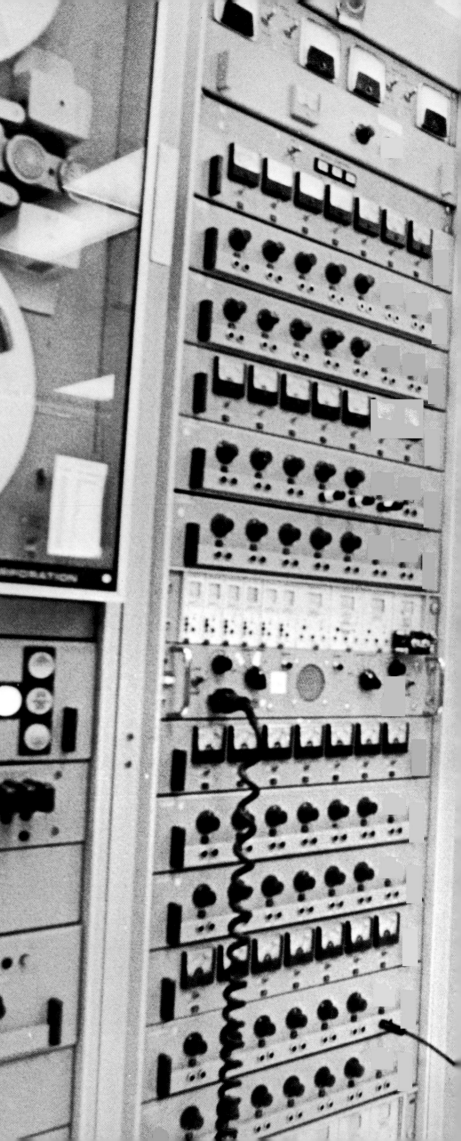

VIKING MISSION

ORBIT:
AREAL MAPPING
THERMAL MAPPING
WATER VAPOR MAPPING
ATMOSPHERIC PRESSURE
PROFILE & ELECTRON
DENSITY

ENTRY:
IONOSPHERIC PROPERTIES
ATMOSPHERIC COMPOSITION
& STRUCTURE

LANDED:
BIOLOGICAL DATA
ORGANIC CHEMISTRY
ATMOSPHERIC COMPOSITION
METEOROLOGICAL PARAMETERS
SEISMIC PROPERTIES
MAGNETIC PROPERTIES (SOIL)
PHYSICAL PROPERTIES (SOIL)

NASA SL71-2525
(Rev. 1) 1-19-72

Above: A diagram of the landing of the Viking Mission spaceship soft landing on Mars.

Opposite: Bound for Mars, Viking 2 was launched 9 September 1975.
Below: An artist's conception of the Mars lander on Mars. The lander — a miniaturized laboratory packed into a hexagonal box 59 inches across and 18 inches high weighed about one ton.

measure the ionic, atomic and molecular composition of the Martian atmosphere. The radiometer would measure surface temperatures. Mariner 6 and 7 also possessed a new type of scan platform which allowed the instruments to be pointed, extending the range of visibility during the flyby. Another innovation was an experimental high-rate telemetry system which enlarged the telemetry capabilities to 16,200 bits per second, 2000 times more than Mariner 4. Data would be gathered in two phases, 'far encounter,' with pictures being taken at three days before flyby and 'near encounter,' which would photograph close-up pictures. This provided for closer resolution between the Earth-based photos of Mars and those taken in near-encounter by Mariner.

Though the flights of Mariner 6 and 7 were successful in fulfilling the objectives of their mission, the data provided was only a brief glimpse of the surface of Mars. To obtain more information would require scores of such flybys. As an alternative, NASA, with the aid of scientists and engineers, planned to place an advanced Mariner in permanent orbit around the planet.

111

Plans for the Mariner-Mars Project began in September of 1968 with a probable launch date scheduled for sometime in 1971. The main objectives for the Project were mapping the planet and observing its dynamic characteristics over a period of three months. The spacecraft would carry a scientific payload of six experiments, similar to those carried by earlier Mariners. There would be television, ultraviolet spectroscopy, infrared spectroscopy, infrared radiometry, occultation and celestial mechanics. A large group of scientists were then selected by NASA. This group was divided into several teams, each representing a specific experiment. In turn, each team would specify to project scientists and engineers the requirements of its particular investigation. Project scientists and engineers would include these specifications as part of the design. It was also necessary to set up a mission-planning phase to determine such things as the best trajectory and the operating sequence.

In the final plan, it was decided that two identical craft would be launched. Their missions would be separate but complimentary. The first spacecraft would be launched as primarily a means of mapping the planet and polar reconnaissance. It would follow an orbit of 11.98 hours with an 80 degree inclination toward the Martian equator and in its closest approach (periapsis) would maintain an altitude of 777 miles (1250 km). It would send back to Earth some 5400 pictures and spectral data during the 90-day orbit.

The second craft would follow a 20.5 hour orbit with an inclination of 50 degrees to the equator and a periapsis of 528 miles (850 km). This, it was felt would allow a high resolution coverage of the mid-altitude range.

Each of the two missions had to be carefully sequenced to insure that each would have full opportunity to achieve its major goals. Each experiment had its own set of objectives. The ultraviolet spectroscopy experiment would study the composition, structure and dynamics of the upper atmosphere of the planet. It would also measure the surface pressure of the atmosphere over most of the planet and search for concentrations of ozone. Like the ultraviolet radiometer, the infrared radiometer was similar to those used on previous Mariners. Its objectives were to measure Martian surface temperatures, checking in particular for areas of irregular cooling rates and the

presence of 'hot spots' which would indicate internal heat activity. It would also detect and measure the minor constituents of the Martian atmosphere, such as water vapor. It would also measure atmospheric temperatures at various altitudes. The occultation experiments would rely on the S-band radio signals. However, with an orbiting satellite, information could be obtained on the shape of the planet as well as variations in the atmosphere relevant to latitude, season, and the time of the Martian day. Celestial mechanics would require no specific instrument other than the spacecraft radio. Its experiments would obtain information to help determine the size, shape, distance and position of the planet.

Mariners 8 and 9 were similar to Mariners 6 and 7. The main difference was that 8 and 9 required larger engines to place them in orbit around the planet. Like all the foregoing Mariners, they were inertially stabilized with their solar panels always facing the Sun. Their roll position was controlled by a star tracker aimed at Canopus. The solar panels were designed to furnish 350 to 500 watts of power to the craft. This power was supplied continuously except

Above: A rendition of Mars made from three separate pictures taken on 17 June 1976 as Viking I closed to within 348,000 miles of the planet.
Below: Viking I obtained this picture of the Martian surface and sky on 24 July 1976. Part of the spacecraft's structure is in the foreground.

Above: A picture of the surface of Mars 21 July 1976 shows that the Martian soil consists mainly of reddish, fine-grained material, with small patches of black or blue-black soil.
Below: Another photo of 21 July. Orange-red materials cover most of the surface.

during propulsion maneuvers when the panels were turned away from the Sun and power was supplied by the spacecraft battery.

The experimental high-rate telemetry system had been especially successful in Mariners 6 and 7. It was therefore standard equipment for Mariners 8 and 9. Scientists, however, anticipated a loss of signal strength late in the mission. This would be due to the increasing distance between Earth and Mars. To compensate for this, they would use lower data rates. So in addition to the initial 16,200 bit-per-second rate, rates of 8100, 4050, 2025 and 1012 bits-per-second were designed in the telemetry system.

To receive and process all this data required extensive ground support facilities. These facilities included tracking systems and computer equipment, not to mention the television cameras and processing equipment for the pictures that would be sent back to Earth. Ground support for the television experiments was quite complex. Activities were divided into two areas called 'real-time' and 'non-real-time' processing.

Real-time processing distributed three versions of each incoming image to science

teams within 24 hours. The television data were transmitted as a set of numbers, each of which represented a light value for the dots that would make up the image. These would be processed by a computer. The first version was called a 'shading corrected' version. The second was called a 'contrast enhanced' version of the first. In this version, the light range had been expanded to pick up more detail. The third version was called a 'high-pass filter.' In this version, the contrast had been even more enhanced to pick up the finest details. These were the immediate photographs that the scientists would examine.

Later, during non-real-time studies, the photos could be subjected to special processes, geometric corrections, photometric corrections, removal of errors or omissions due to telemetry and the residual images known as 'ghosts.' Later, after these pictures had been enhanced, rectified and scaled, they were assembled into large maps of the vast Martian surface.

Nearly two years of planning and designing had gone into the launching of these two spacecraft. Early in 1971 they were brought from the Jet Propulsion Laboratory to Kennedy Space Center. In careful sequence, each craft was prepared, fueled, enshrouded and mated with its launch vehicle. On 9 May Mariner 8 was launched into the night sky. The elation of those who watched the launch turned to bitter disappointment. A malfunction in the guidance system of the Centaur rocket, caused the craft to tumble. As it did so, the payload separated and the engine was cut off. Mariner 8 crashed into the sea 350 miles (560 km) northwest of Puerto Rico.

Disappointing as the loss of Mariner 8 was, NASA still had Mariner 9. Project personnel set to work to redesign the Mariner 9 mission so that it would incorporate the major objectives of both missions. Twenty-one days later, Mariner 9 was successfully launched. Six days into its voyage to Mars, a trajectory correction was accomplished so accurately that no other corrections were needed during the rest of the 167 day trip to Mars. Mariner 9 traveled 245 million miles (394,000,000 km) and arrived at its destination within 31 miles (50 km) of its target point.

Because the planners of the Mariner 8 and 9 Mission decided early to maintain an 'adaptive operational capability' this meant that the mission plan could be changed on short notice. Thus it was that Mariner 9 was

Viking Orbiter 1 took this picture from 19,200 miles above Mars. The equator cuts across the top left corner.

able to achieve every goal which had been outlined for the two spacecraft. Over 7000 pictures were sent back to Earth and the entire surface of the planet was mapped. The success of Mariner 9 was truly phenomenal.

But did life actually exist on Mars? The observations of Mariner 9 did not prove or disprove the possibility. If it did exist, it would have to survive in an atmosphere of carbon dioxide, an atmosphere so thin that it would equal that of the Earth 20 miles (32 km) above the surface. It would have to survive the strong radiation of the sun, particularly ultraviolet radiation. It would have to survive sudden and extreme temperature changes ranging from a balmy 80 degrees Farenheit (27°C) at noon-day on the equator to a chilling 150 degrees below zero (−65°C) at night. It would have to get along with little or no water. No water in liquid form appeared to exist on Mars, despite the appearance of erosion and dry 'river beds.'

In the midst of all the questions, the Viking program came into being. Nearly ten years of planning and complex designs went into the Viking Project. Then on 20 August and 9 September 1975, the twin spacecraft were launched. They would cross 400 million miles (643,000,000 km) through the empty darkness of space to reach their goal, the 'red' planet, Mars. It would be almost a year before they reached their destination.

Each craft consisted of two parts: an orbiting satellite which would photograph the Martian surface and a lander which would separate from the satellite and descend to the planet's surface. As it came closer to the chosen landing site, retro-rockets would be fired to slow its descent. Parachutes would be deployed, slowing the descent further. Viking I began its orbit around Mars on 19 June 1976. Its task was to select a landing site for the lander. Three Russian probes had previously been des-troyed on impact with the Martian surface. Selecting the site was, therefore, no mean task. Over a billion dollars had gone into the Viking Project and NASA was not anxious to see the Viking lander come to such a fate.

Out of cautiousness, the site that seemed most desirable would be smooth and fairly level. The data collected from the orbiter cameras and from the Earth-based radar scans would be conducted from the largest radiotelescope on Earth at Arecibo, Puerto Rico. Several possible sites were rejected as

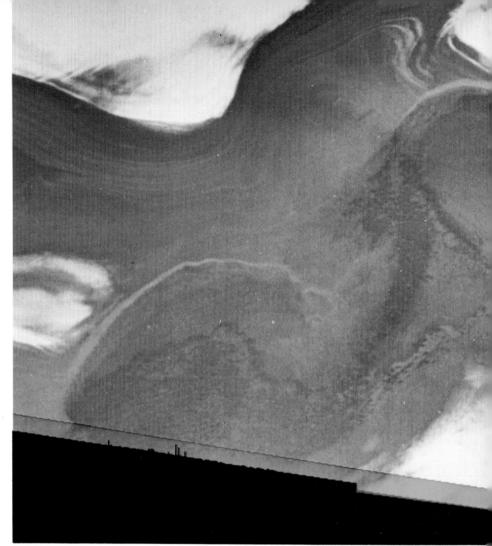

A frosty scene near Mars' north pole shows the region in midsummer when the seasonal carbon-dioxide polar cap clears to reveal water ice and layered terrain beneath. The picture was taken from a distance of 1370 miles.

Below: An artist's rendi-tion of Olympus Mons on Mars taken from a NASA film.

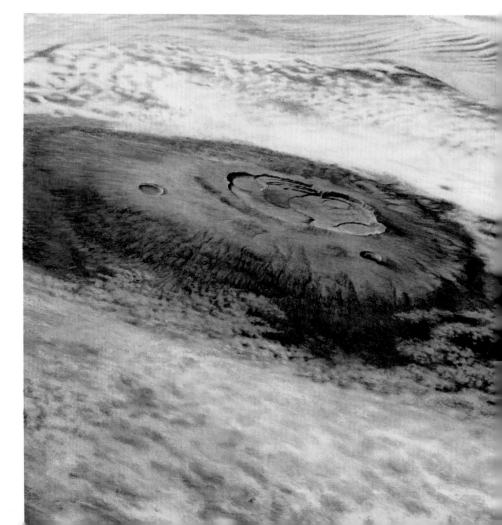

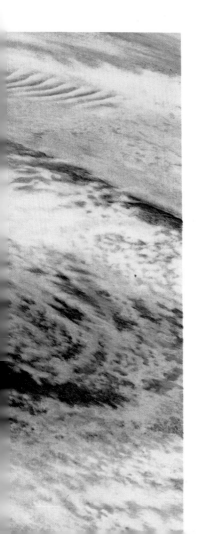

'too rough' before the site at Chryse Planitia (Plain of Gold) was chosen. On 20 July 1976, Viking I landed successfully on Mars. Within an hour of the landing, the first photos of Mars were being radioed back to Earth. Viking scientists were jubilant.

Poised on its three metal legs, the Viking lander began to function by activating all of its electronic sensors. It rotated its twin cameras, capable of giving a three-dimensional view of the landscape, to focus on the horizon. The mass spectrometer tested the atmosphere to determine its composition. A metal arm rose up to check the weather conditions, temperature, barometric pressure, wind speed and direction. All of this data was relayed to the Viking orbiter, which in turn relayed it back to Earth. In size, the Viking lander was similar to a compact car. Literally all of its equipment, the cameras, the transmitters, sampler 'arm' and over a dozen test devices and sensors were powered by a single 50 watt generator. This amounts to less power than that used by an average household light bulb.

Within the limited confines of the lander were three automated chemical laboratories. These contained tiny ovens, filters, counters for radioactive tracers, a gas chromatograph which would identify chemical substances and a lamp which would reproduce the weak sunlight found on Mars. All of this was packed into one cubic foot! Along with all of this were 300,000 transistors, 2000 other electronic parts, 1000 wire connections and 37 miniature valves. The miracle is that all of these functioned perfectly despite the great distances traveled, the ordeal of landing and the harsh climate of Mars. (The range of temperatures experienced on that first Martian day went from −23 to −123 degrees Fahrenheit (−30 to −86 °C.)

Most of the tests were aimed at discovering life or some evidence that life had existed on Mars. One cell contained a mixture that scientists dubbed 'chicken soup.' Nutrient-rich, it was supposed to encourage growth, reproduction or metabolization of whatever life form might exist on the planet. This was all accomplished with extremely complicated machinery. A delicate-looking steel arm with a shovel at its end would reach out to a distance of ten feet. At that point, it would turn at the wrist and drop to the ground. When the shovel made contact with the ground, it would click open. It would then push firmly into the soil, scooping up a sample and click closed. Then it would swivel its wrist again and retract the long arm. The precious load of soil was carried back to the spindly little craft. Over a cylinder covered by a wire screen, the arm paused. The shovel lid began vibrating violently. Dust and rock particles sifted from holes in the shovel. The earth samples disappeared into the craft. Within the craft, measured samples were carried by a rotating conveyor to several test cells. Each sample would receive a battery of tests, designed specifically to find evidence of life.

Even with all this complex machinery, all the careful tests, scientists to this day cannot say with certainty whether there is life on Mars. Some of the experiments such as the 'labeled release' test produced very strong evidence for the presence of some type of microbe similar to Earth bacteria or algae. In this test, soil was moistened with the nutrient solution. Then the amount of carbon-dioxide gas was measured as it was discharged from the soil. The amount of carbon dioxide released in this experiment was quite high; exactly what would be expected from the metabolism of microorganisms. The results for this test were the same at each of the two Viking lander locations. Each control test, in which the soil was first sterilized by heat, showed no release of any gas. This test alone would seem almost positive proof of living organisms on Mars. But the reactions from other tests were so strange that scientists declined to make any positive statements.

One of the other life-detection tests called the Gas-Exchange Experiment produced a sudden and unexplainable burst of oxygen when Martian soil was exposed to moisture. Scientists felt this seemed to be a non-living chemical reaction rather than a response produced by living organisms. Perhaps the 'labeled release' test was also a chemical reaction. The biggest puzzle of all was the results of the GCMS (gas chromatograph-mass spectrometer). The object of this experiment was to find organic (carbon-based) chemicals in the Martian soil. Carbons occur naturally, even in the depths of space. Carbon is one of the compounds common to all life. Yet the Martian soil yielded none of these organic chemicals. It was baffling. It is still baffling, although some recent tests done with Antarctic soil show much the same reactions as those done on Martian soil. Antarctica comes close to duplicating the temperatures and lack of moisture on Mars. Is there life on Mars? The answer is still a calculated 'maybe.'

Ambassadors
to the Universe

Perhaps the most successful and amazing of NASA's unmanned spacecraft were the Pioneers. Weighing a mere 570 pounds (259 kg) and looking a little like a radar dish with wings, the Pioneer 10 was the first craft to leave our solar system and cross into deep space. It carried an ultraviolet photometer, an imaging photopolarimeter, a geiger tube telescope, a meteoroid detector sensor panel, helium vector magnetometer, plasma analyzer, a trapped radiation detector, a cosmic ray telescope, infrared radiometer and a charged particle instrument.

A plasma of electrically charged particles exists in interplanetary space, spreading out from the Sun across the solar system. Magnetic fields permeate this space. While scientists had observed and measured the fields up to the orbit of Mars, they knew little of what lay in the outer regions of the Solar System. Pioneer 10 and 11 would map the magnetic field of interplanetary space. They would measure particles, fields and radiation.

The complexities involved in exploring deep space were mindboggling. First of all, it would be necessary for the spacecraft to leave Earth's orbit under very high velocity to reach its destination in a reasonable amount of time. Therefore, Pioneer 10 and 11 were launched at the highest velocity ever achieved by a man-made object. Using an Atlas-Centaur launch vehicle, the spacecraft blasted off at speeds of 32,000 miles per hour (51,500 km per hour). This boosted each spacecraft into a direct ascent with none of the usual parking orbit around the Earth. The shortest possible time to Jupiter would be just under 600 days, considering the capabilities of the launch vehicle. The longest trip would involve 748 days. There would be several inflight maneuvers made during the Pioneer 10 mission. These would place it on a target with Jupiter at a time and position best suited to observe the planet as well as some of its larger satellites. In-flight maneuvers for Pioneer 11 would include the option of continuing on to Saturn.

Pioneers 10 and 11 would utilize the already-proven modules of Pioneers 6–9. The craft would be small, lightweight and magnetically clean. In addition to the

Previous spread: A montage of images of the Saturnian system prepared by an artist from an array taken by the Voyager 1 spacecraft during its Saturn encounter in November 1980.

Atlas-Centaur launch vehicle, there would be a third solid-propellant stage. The craft would be compatible with the launch vehicle. Their communications systems would be compatible with the Deep Space Network. The environment for the scientific instruments had to be thermally controlled. And the basic design must be one that would continue to function in space for many years. Each craft would carry a data system that would monitor the scientific instruments and transmit scientific and engineering information back to Earth regarding the condition of the craft and its instruments. Further, the spacecraft had to be capable of being commanded from Earth to perform their tasks. They must also be able to change the operating modes of the onboard equipment.

The Pioneers would travel in a curved path toward their destination, Jupiter. The distance they would travel would be about 1000 million kilometers or 620 million miles. The curved path would cover 160 degrees azimuth around the Sun traveling between the orbits of the Earth and Jupiter. The Earth would make almost two orbits of the Sun during the flight, while Jupiter would complete only one-sixth of its solar orbit.

A great deal of planning went into all of the unmanned flights. In the case of the Pioneers, there were some options available as to flight path and launch dates. Timing would be very important. Should one of the craft arrive at its destination too early, the sensors on the spacecraft could not perform the desired scientific experiments. Other arrival dates would have conflicted with the flight of Mariner 10. The huge 210-foot (64-m) antennas of the Deep Space Network could only receive one set of signals.

Pioneer 10 could have been launched anywhere between 25 February and 20 March 1972. The flight would then have arrived at Jupiter sometime between mid-October 1973 and late July 1974. The arrival time would have to be scheduled so that Jupiter and the spacecraft would not be too near the Sun when viewed from the Earth. Both the orbit of the Earth and the path of the spacecraft would be considered carefully. At about 300–325 days and 700–725 days after the launch, the Earth

Left: An artist's concept of Pioneer over the Jovian surface. Jupiter appears to have its own energy source and is so massive that it is almost the size of a small star.

and the spacecraft would be on opposite sides of the Sun. For Pioneer 10 to arrive at Jupiter more than 700 days after launch would not be practical. A further point to be considered was that when the spacecraft passed behind the sun, communications would be interrupted. This communications blackout should not occur at some crucial point in the mission.

There was then the problem of targeting. How close should the spacecraft be allowed to approach Jupiter? How should the trajectory be inclined to the planet's equatorial plane? What should be the point of the closest approach to the equatorial plane? In planning the trajectory, scientists were concerned with the possible radiation damage that might occur. In the end, they decided to risk damage in order to obtain the maximum amount of information. Thus the image of Jupiter could only be obtained before the closest approach. The approach trajectory that was finally decided upon would show the planet well-illuminated just before actual encounter. Jupiter would then appear as a partially illuminated crescent just after the encounter.

Scientists had first thought that occultation of the planet Jupiter should be avoided. But there would be no other feasible way to study the Jovian atmosphere, so an occultation was selected.

A kind of gravity slingshot effect which would serve to 'bounce' Pioneer 10 into deep space was under consideration. The question was, how close could the spacecraft approach Jupiter without suffering damage to its optical and electronic equipment? Jupiter has strong radiation belts trapped within its magnetic field. How would this radiation affect the functioning of the craft? No one could say with any certainty.

In July 1971, a group of scientists at the Jet Propulsion Laboratory analyzed all of the available information about Jupiter. From this, they defined the probable environment of the planet. Again, no one could be sure that this was the actual environment. Much would have to be left to the mission to discover. But nevertheless the design of the Pioneer spacecraft and the instruments it carried were based on this presumed environment.

Radiation was not the only hazard of the mission. There are perhaps 40,000 to 100,000 small asteroids which orbit our solar system. Many are found between the orbits of Mars and Jupiter. Some measure

several hundred miles in diameter, others are much smaller. They may travel at speeds of 60,000 mph (97,000 km/h). The orbits of the larger asteroids are known but many of the smaller ones are unknown. If Pioneer 10 were to collide with even the tiniest of these asteroids, it might be seriously damaged. If further deep space missions were to succeed, at least one spacecraft had to penetrate the asteroid belt and survive.

Scientists faced another problem with the Pioneer mission. How could electrical power be supplied to the spacecraft at such a distance from the Sun? At first solar cells had been considered as a source of electric power. But sunlight at Jupiter carries only 1/27 of the energy that it has on Earth. It would therefore be necessary to carry a massive array of solar cells. This presented the problem of the Jovian radiation belts again. Radiation could damage the solar

Above: A computer operator (background) and a programmer input instructions to the Ames Research Center's Sigma 5 computer for use in the Jupiter Mission. It took 46 minutes for the data to be transmitted between Earth and Jupiter.
Opposite: A stylized view of Pioneer leaving Earth and passing the moon. The trip to Jupiter lasted almost two years. Jupiter has more than twice the mass of all the other planets combined.

cells seriously. The solution for the problem of electrical power turned out to be radio-isotope thermoelectric generators.

Tremendous speed would be required to carry the spacecraft from the Earth to Jupiter. The payload would have to be extremely restricted. There would be no complicated onboard computers. These would be much too weighty. The Pioneer would be directed from the Earth despite the extensive delays in communicating over such vast distances.

The watchwords in all things connected with the Pioneer spacecraft were reliability and simplicity. Vital items such as transmitters and receivers were duplicated. Only systems and components that had proven reliable on other spacecraft were used. Every electronic component was rigorously tested before it was even assembled on the spacecraft. Thus any component which might fail was eliminated before the flight. Much of the success of the mission would depend on the advanced command, control and communications systems linked to computers and human controllers who guided the spacecraft from Earth.

Once the actual mission got underway, there were five distinct phases of operations. Each would require different approaches and techniques. From the standpoint of critical timing, the Earth launch and the planetary encounter were the most stressful. In these situations, the controllers would have to correct quickly any problems which arose. There would be no time for mulling over a decision—every second was vital. During the three interplanetary phases, however, time was less critical.

Various teams participated in each phase. During the prelaunch and launch at John F Kennedy Center the team from Ames Research Center controlled the spacecraft while the team from Lewis Research Center controlled the launch vehicle. After separation of the spacecraft from the launch vehicle and entrance into the transfer orbit to Jupiter, control of the spacecraft went to the Ames flight operations team at Jet Propulsion Laboratory.

At the same time, control of the scientific instruments within the spacecraft was handled by the Pioneer Mission Operations Center at Ames Research Center. By arranging this period of split control, the mission could take advantage of the multiple consoles and computers at the Jet Propulsion Laboratory during the first critical days of the flight. It also allowed the engineers to

monitor all the subsystems; telemetry, power, thermal, attitude control, data handling and command.

Eleven hours after liftoff, Pioneer 10 passed the orbit of the moon. (Compare this with three days for the Apollo spacecraft to reach the moon.) At this point, the monitoring activities began to change. Previously, scientists and engineers had been involved with the functioning of equipment on board the spacecraft. Now concern shifted to readying the craft for its long voyage to Jupiter and beyond. A few days after liftoff midcourse maneuvers set the flight on course for Jupiter. All equipment and scientific instruments were performing well. Now, control shifted once again. Mission crews left the Jet Propulsion Laboratory and the Kennedy Space Center to return to Ames Research Center which would now become the control center for the entire mission.

This was now Phase Three, the interplanetary mode. For those monitoring the flight, the task was one of watching and waiting. The delays were unavoidable and increasing; it was a measure of the distance Pioneer 10 had traveled. Signals traveled from Earth to the spacecraft and back from the spacecraft to Earth. During this phase, a team of from five to seven scientists with a support team from Pioneer Missions Operations Center monitored the spacecraft. The incoming data was continually monitored by computers and personnel. In the event of a malfunction, ground control could be alerted at the earliest possible moment. This would be especially important if corrective measures were needed.

At Ames Research Center a computer monitored telemetry signals on both Pioneer 10 and 11, checking the functioning of the spacecraft and their payloads. Had a voltage or a temperature raised or lowered, or any status of an instrument changed without having been commanded to do so, the computer would sound an alarm. This would be followed by a printed-out message. At this point, the duty operator would notify the proper engineer or scientist who would then resolve the problem. Each of the mission controllers had a list of specific procedures to cover any emergency. They had also been advised where they might obtain technical help if needed.

During the voyage, scientists and engineers sampled data from each scientific instrument. This was done to determine how well the instruments were functioning, scientifically as well as in the engineering sense. Scientists, engineers and controllers watched for the need to change bias voltages, to switch modes of operations, to adjust the range of sensitivity of the instruments.

As the Pioneers reached the edge of Jupiter's atmosphere, quick action again became necessary. But, at this point, the action was slowed by the distance. The tiny spacecraft were, at this point, over 500 million miles (800,000,000 km) from Earth. Radio signals took 92 minutes to travel from Earth to the spacecraft and back again. (When Pioneer 11 reached Saturn the radio signals took 173 minutes round-trip.) It was therefore necessary to have all commands well planned in advance because of the delay.

The imaging photopolarimeter (IPP) became a very critical piece of equipment at this point. It could be given a long sequence

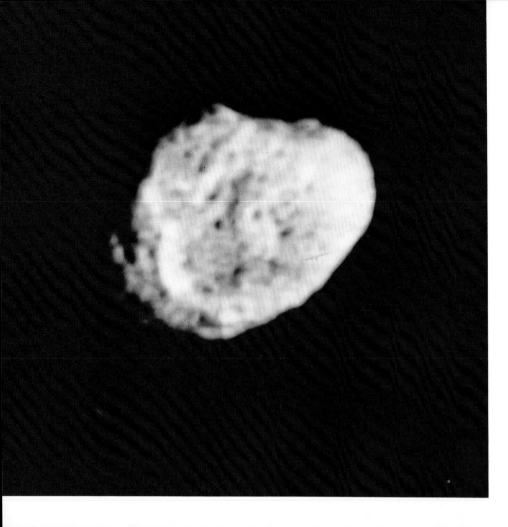

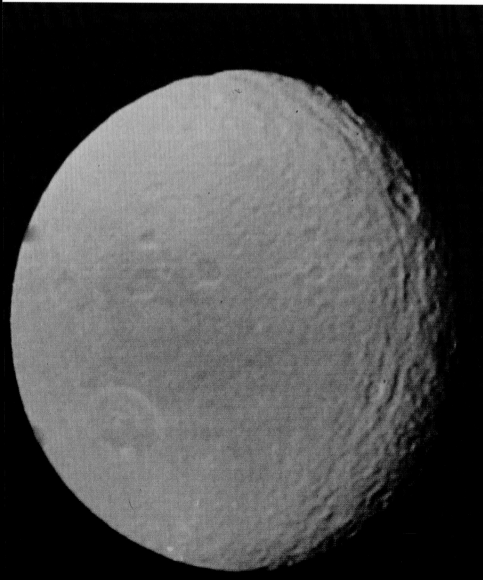

of commands which could be utilized as the spacecraft passed by the planet. Should there be an accumulation of electrical charges resulting in spurious commands, there was a set of contingency commands which would reconfigure the spacecraft and its instruments. This was the fourth phase in the mission.

As the Pioneers passed beyond the planets, Jupiter and Saturn, they entered the final phase of the mission command and control. Traveling farther and farther beyond the planets, their signals grew fainter and fainter. In their travels, they would be tracked by the NASA Communications Network, operated by Goddard Space Flight Center. The Goddard Space Flight Center sponsored worldwide ground communications circuits and facilities, linking all the Earth terminals receiving signals from the spacecraft with the control centers on the west coast of the United States. Further communications came from the Deep Space Network which was operated for NASA by the Jet Propulsion Laboratory and which extended around the world. The Deep Space Network provided deep space telemetry data acquisition as well as tracking and commanding capabilities through its 85 foot (26 m) and 210 foot (65 m) diameter antennas. These were located at Goldstone, California, Spain, South Africa (until 1974) and Australia. As the Earth turned on its axis, these giant antennas tracked the distant journey of the Pioneers. In some cases, two stations might have an overlapping coverage. Such was the case with the encounter with Saturn. Both Madrid and the Goldstone stations were receiving signals at the same time. This was a critical period. The fact that both stations were receiving signals reduced the chance of losing vital data.

Because travel over such vast distances raised communications problems never encountered before, NASA relied heavily on the 210 foot (64 m) antennas of the Deep Space Network. Since transmitters and antennas on board the spacecraft had to be extremely lightweight and had to conserve power, the super-sensitive large antennas

Top left: A Voyager 2 view of Saturn's satellite Hyperion at a distance of about 300,000 miles. The irregular shape is probably the result of repeated impacts that have taken off large pieces. Left: Tethys, another satellite of Saturn, taken at a distance of about 368,000 miles. Previous spread: The Jet Propulsion Laboratory's control center of the Deep Space Network.

like that at the Goldstone Station were especially important. The 85 foot (26 m) diameter antennas could be used when the larger antennas were required for other space missions. But they received information at a much lower rate.

The larger antennas could direct their radio beams so precisely and provide such a high radiated power (Goldstone could produce up to 400,000 W) that commands could be received by the spacecraft at distances up to several hundred times the distance between the Earth and the Sun. Compare this with messages which were sent from the spacecraft which could be received on Earth at a distance of perhaps 40 times that between the Earth and the Sun!

The spacecraft itself carried three antennas to receive and send signals. There was a low gain (omnidirectional) antenna, a medium gain and a high gain (beamed) antenna. It also carried two (redundant) receivers for commands from Earth and two transmitters which would send signals back to Earth. The dual transmitters and receivers provided a back-up in case of failure during the long journey.

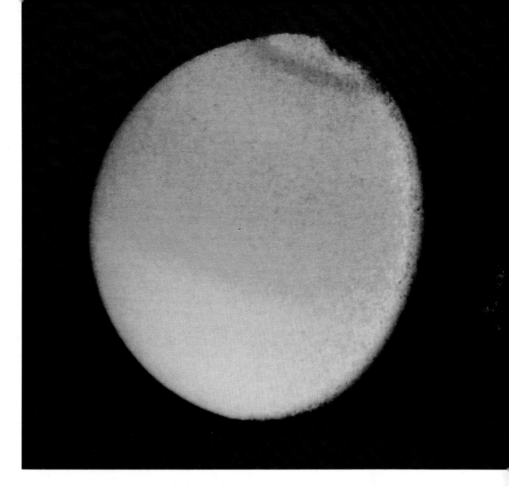

Above: A Voyager 2 photograph of still another moon of Saturn - - Titan - - from 1.4 million miles.
Below: A montage prepared from images taken by Voyager 1. Six of Saturn's satellites from the right are Tethys, Mimas, Enceladus, Dione, Rhea and Titan.

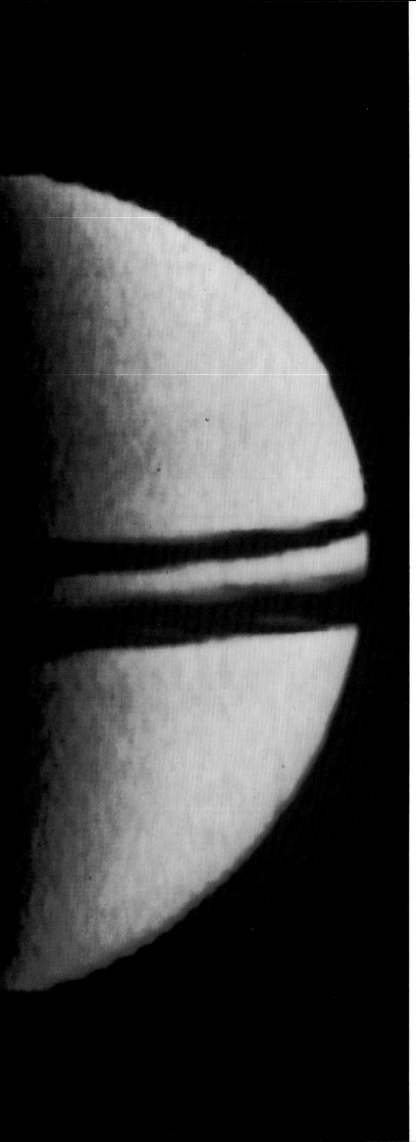

On Earth, the amount of energy received from the Pioneers via radio signals from Jupiter was incredibly small. From the distance of Saturn, the amount is smaller by more than two-thirds. One can get an idea of how infinitesimal that is by realizing that an 85 foot (26 m) diameter antenna, collecting this energy from the distance of Jupiter, would need 17 million years to collect enough energy to light a 7.5 W nightlight for one thousandth of a second! Because of highly sophisticated data coding and signal modulation techniques, because of the large, ultra-sensitive antennas, and highly advanced ultra-cold receiving devices attached to them, it was possible to receive these distant signals.

Some 11 years and 2.8 billion miles (4.5 billion km) from its launch site, signals from Pioneer 10 were still coming back to Earth. Signals from the 8 watt transmitter reached Earth as a series of electronic beeps, almost musical in tone. At that great distance, radio signals take nearly 4 hours and 20 minutes to reach Earth. The signal that reached the Earth amounted to just 20 trillionths of a watt. At 64 bits per second they were inaudible. The signals had to be amplified by computers and slowed to 8 bits per second. Much of what was transmitted at this point was scientific data. Another 30 percent was steering data. The instruments used in the exploration of Jupiter are no longer turned on. Much of the material transmitted by this distant spacecraft will take years to analyze.

Pioneer 10 was the first spacecraft to leave our solar system. Alan Fernquist, assistant flight director at NASA's Ames Research Center in Mountain View, California, feels the 570 pounds (26 kg) spacecraft will probably last forever. Over the next million years, it will not encounter any other objects. The closest approach to any star will be in 32,610 years when the tiny spaceship passes within 3.3 light years of the red dwarf, Ross 248. What has been most astonishing is the continued performance of Pioneer 10 year after year. Engineers at TRW where the Pioneer was created, have been amazed that it continued to survive the harsh conditions of deep space.

Somewhere in the eternal darkness that is outer space, a rather curious-looking pair of spacecraft hurtle toward a rendezvous with the outermost portions of the solar system. Project Voyager began as a dual spacecraft mission. The spacecraft were

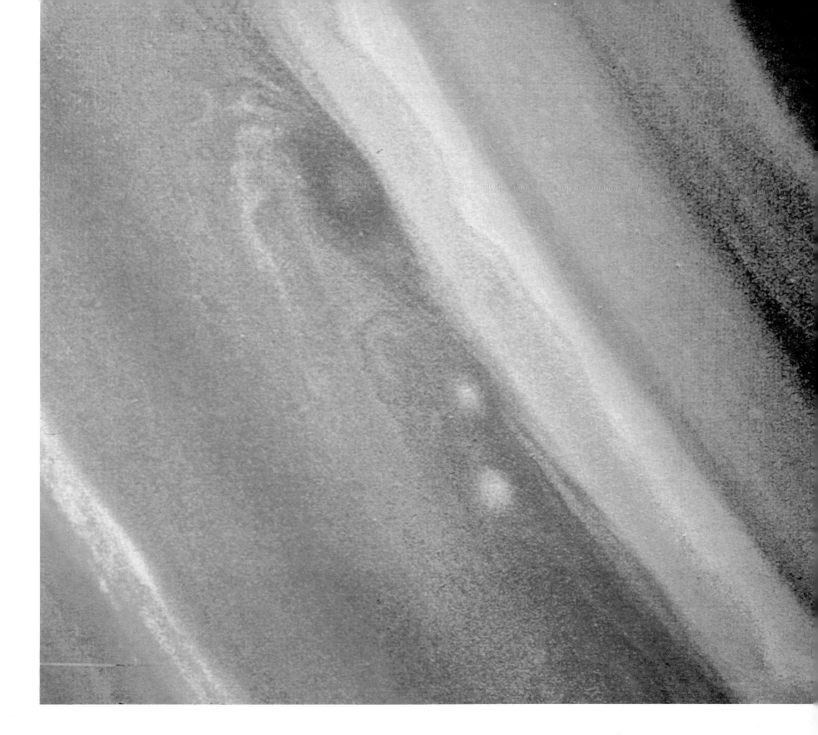

launched in August and September 1977 with the immediate goal being to encounter the four planets of the outer solar system. While the first four planets of the inner solar system—Mercury, Venus, Earth and Mars—are relatively similar in form, the four planets of the outer solar system are quite different. So different, in fact, that they seem to be more akin to stars than to planets. It was to explore and send back data on these mysterious planets, Jupiter, Saturn, Uranus and Neptune, that the Voyagers were launched. Jupiter was encountered early in 1979 by Voyager 1. This, despite the earlier Pioneer flights, was the first close encounter with the huge planet. In November 1980, Voyager 1 encountered Saturn while a second look was afforded by Voyager 2 in August 1981.

Each Voyager weighed 1797 pounds (815 kg). The central part of each spacecraft consisted of a ten-sided bus which consisted of a 54 pound (24.5 kg) aluminum framework ring measuring 18 inches (45 cm) high and 70 inches (179 cm) across. Within the ring were 10 electronic packaging compartments. The ring surrounded a titanium sphere, the fuel tank, which at launch, contained 229 pounds (104 kg) of hydrazine propellant for the 16 thrusters. Above the bus was the high-gain antenna, probably the most outstanding feature of the spacecraft. The high-gain antenna was a large saucer that transmitted and received on two frequencies, the S and X bands. The low-gain antenna, which was used during the first 80 days of flight, was supported over the high-gain dish. Each Voyager was

Above: Saturn's northern hemisphere from Voyager 2 at a distance of 4.4 million miles.
Opposite: A Pioneer computer-enhanced picture of Saturn.
Previous spread: An artist's concept shows the Voyager 1 spacecraft as it passes through the flux tube of Io, a region of magnetic and plasma interaction between this satellite and Jupiter.

133

powered by three radioisotope thermoelectric generators (RTG) located on the boom. These were deployed one hour after launch.

The scientific equipment carried by each spacecraft consisted of 10 experiment packages and a radio transmitter located on the scan platform boom. These were deployed 180 degrees from the generator boom to avoid possible radiation damage from the generator.

The optical scanners included a pair of TV cameras plus four particle and field detectors which would measure interplanetary plasmas, low-energy charged particles, cosmic rays and magnetic fields. Each craft also carried four magnetometers, two high-field magnetometers located on the bus and two low field located at the end and mid-point of a long 45 foot (13 m) boom. The remaining scientific instruments package was a pair of 33 foot (10 m) antennas for studying radio emissions.

In addition to the scientific equipment, the Voyagers each carried a unique cultural message which described life on the Earth and the inhabitants. Though the chances of encounter with extra-terrestrial civilizations are extremely small, it is possible that millions of years in the future these craft may be found by aliens from another world.

Once their explorations are complete, the Voyagers will slowly leave the solar system. Thus, each carries with it a gold-coated copper phonograph record. Contained on the record are 90 minutes of the world's great music, greetings in 60 languages (and one whale language!) an audio essay entitled 'The Sounds of Earth,' and 118 photographs of the Earth and its civilization.

Much went into composing this record. How, for instance, would earth people communicate with those from another planet? In a world where nothing is in any way similar to that which we know, mathematical relationships should still be valid. A simple binary code using two numbers, 1 and 0, was established. In the 1271-character message, the numbers can be decoded into a kind of picture showing a man, woman, child, and the solar system.

Right: The 'Sounds of Earth' recording is mounted on the Voyager 2 spacecraft—4 August 1977.
Below: The 'Sounds of Earth' recording. The record, made of copper and gold-plated, contained two hours of sound and music, and digital data including pictures and a message from Jimmy Carter.

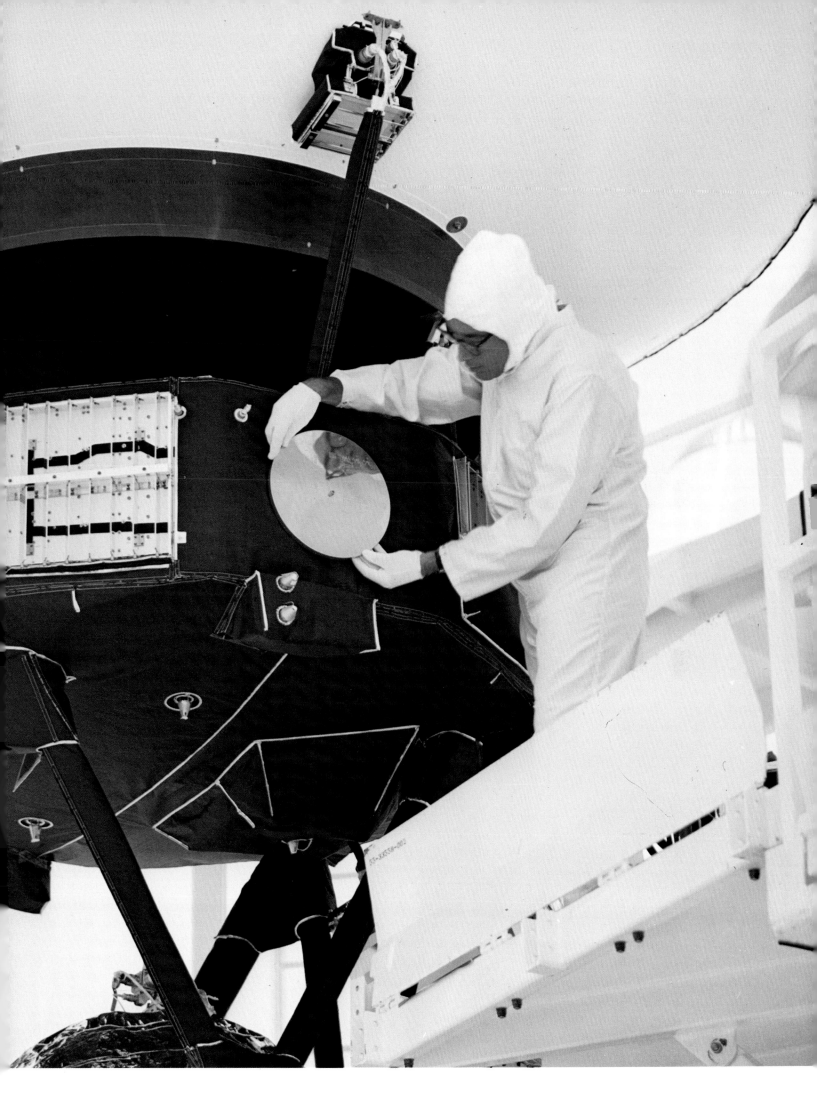

Another portion of the record is devoted to music from around the world. Included were Bach's Brandenburg Concerto No. 2, 'Kinds of Flowers' from Java, Senegal percussion, songs from Zaire, Australia, Mexico, New Guinea, Japan, The 'Queen of the Night' Aria from Mozart's *Magic Flute*, 'Tchakrulo' from Russia, 'Melancholy Blues,' by Louis Armstrong, a Navajo Indian night chant and Beethoven's Fifth Symphony, among many other works from around the world.

Since this particular portion of the Voyager Project was put together in a very short time—just six weeks—there was actually only a month to select everything that would go into the record. All that went into the record had to be stated in a formal proposal. All the pictures, voices, music and sounds had to be included in the proposal. Not only that, but also the specially approved and fabricated record had to be ready to bolt on to the Voyager. Within one month's time, all the photographs had to be found and legal permission for their use had to be secured. All the diagrams had to be drawn and a means of converting the pictures to a sound signal so that they could be recorded had to be found. And a diagram explaining how to play the record had to be devised.

Recording the pictures threatened to become the greatest problem. NASA was not prepared to lay out large sums of money for this endeavor. Luckily, Frank Drake and Valentin Boriakoff, a research associate at the National Astronomy and Ionosphere Center, found Colorado Video. The company allowed the use of their equipment to convert the pictures into sound. The next question was, what pictures should be sent? It was decided not to send pictures that presented war, disease, crime or poverty. The records which would travel with the Voyagers might well be the only surviving images of the Earth. It seemed best to send only those pictures which would represent mankind's highest ideals.

Another thought which occurred to those who assembled the record was whether or not the extraterrestials would be able to understand pictures. Perhaps they would not have the same senses that human beings have. Some of the pictures chosen were therefore very simple. The first two pictures showed objects which were found elsewhere on Voyager. They were of the circle and the pulsar map found on the record's cover. Other pictures show the Earth as viewed from space, the moon, the structure of living

cells, the skeleton and muscles of the human body, a developing fetus, parents and children, the ocean, a snow-topped mountain peak, a seashell, insects, trees, fish, animals, houses, highways, factories, musical instruments and an astronaut. Some of the pictures were in color; some were in black and white.

The greetings range in sentiment from highly formal to charmingly personal. One in Mandarin Chinese sounds almost like a postcard one would send to a vacationing friend. It says, 'Hope everyone's well. We are thinking about you all. Please come here to visit us when you have time.'

One in Latin says, 'Greetings to you, whoever you are: we have good will towards you and bring peace across space.'

An Indian, speaking in Gujurati, sent this message: 'Greetings from a human being of the Earth. Please contact.'

Messages were recorded from populations all over the world. Since there were only two recording sessions for the greetings, both done at Cornell University, a great deal of time went into locating people fluent in the various languages. Despite the

tight schedule, a remarkable number of languages are represented.

Those who watched the Voyagers' launch must have felt much like the people who lined the docks in 1492, watching Columbus sail across the sea in three tiny ships. What lies out there, beyond our solar system? It may be a hundred thousand years before we have a reply. It may be never. But the message has gone out and is even now traveling toward the stars.

More immediate returns have come back to Earth on the distant planets, Jupiter and Saturn. Jupiter, which was the first of the planets encountered by Voyager, has a mass 318 times that of Earth. Thirteen hundred Earths could fit in the volume of Jupiter. For all its size, Jupiter has only one quarter the density of the Earth. Most of its composition is of hydrogen and helium in massive quantities. Scientists believe that when the solar system was formed, 70 percent of the available material was taken up by Jupiter. As a huge swirling cloud of dust and gas it suddenly collapsed into a huge, glowing, red ball, perhaps 124,300 miles (200,000 km) wide. Since then, the planet has shrunk

Above: An artist's concept of the flight of the unmanned Voyager spacecraft as it passed Saturn. Note the large dish antenna for long-distance communications with the Earth. The three circular canisters at the bottom are the nuclear-electricity power generators.
Opposite: A drawing of Pioneers F and G, built to investigate the interplanetary medium and the nature of the Asteroid Belt, and to survey Jupiter and its environment.

to a mere 88,900 miles (143,000 km) in width. A day on Jupiter is only ten earth-hours long. But because it orbits some 448 million miles (778,000,000 km) from the sun, a year on Jupiter is equal to 12 Earth years.

Somewhere beneath the clouds which are 62 miles (100 km) deep, lies a core of silicates and metals, probably about the size of Earth. But with its main composition being hydrogen and helium, Jupiter is much more like the sun. Temperatures at the cloud level are fiercely cold—minus 184 degrees Fahrenheit ($-120°C$). Temperatures at the center may be as high as 54,000 degrees Farenheit ($30,000°C$)—five times the temperature of the sun's surface. Near the center of the planet the pressure equals 100 million Earth atmospheres.

The results of such pressure turns the hydrogen to a liquid and also converts it to a metallic form. This metallic hydrogen is an electrical conductor. Combined with the internal heat, it creates an enormous electrical field around the planet which extends for a distance of 4.35 million miles (7,000,000 km). Trillions of high energy particles are trapped within the invisible radiation belts. This region is called the magnetosphere.

At times, some of the trapped particles are released from the magnetic field. They are detected as cosmic rays as far distant as the planet Mercury, some 435 million miles (700,000,000 km) away. All of Jupiter's 14 satellites orbit within the magnetosphere. They are bombarded constantly by high energy electrons, protons and other deadly ions. The only place that might be safe for a manned spacecraft landing would be Callisto, 1,170,000 miles (1,880,000 km) distant from Jupiter. Io, a mere 262,000 miles (421,600 km) distant, suffers erosion from the severe radiation. In addition, as Jupiter's magnetic field passes Io, an electric

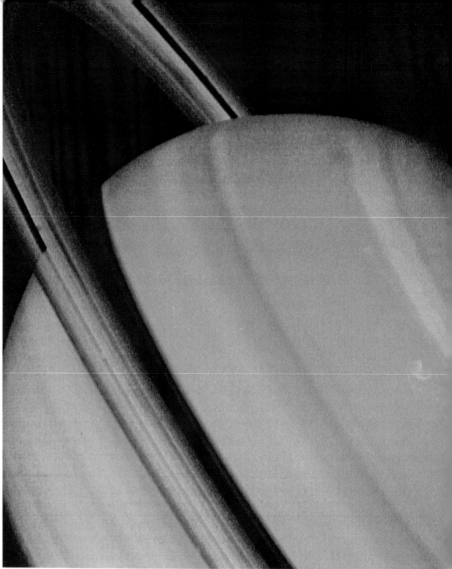

Above: A picture assembled from Voyager 2 Saturn images from a distance of 13 million miles. Three of Saturn's icy moons are visible at the left—Tethys, Dione and Rhea. The shadow of Tethys appears on Saturn's southern hemisphere.
Above right: Voyager 2 returned this view of Saturn and its rings when it was 8.6 million miles away and approaching the planet at about 620,000 miles per day.
Opposite top: A view focusing on Saturn's C-ring (with the B-ring at top and left), made from 1.7 million miles away. More than 60 bright and dark ringlets are evident.
Opposite bottom: A Voyager 2 image shows the region from 20 north latitude to Saturn's Polar Region.

current is generated between the planet and its moon. This current was measured by Voyager 1 at five million amperes. Five million amperes translates into two trillion watts. Two trillion watts is the capacity of all the power plants on Earth. It is perhaps this as much as anything that accounts for the fact that active volcanoes exist on Io.

Io is further subjected to the gravitational pull of not only Jupiter, but also its sister satellites, Ganymede and Europa. These forces heat the upper layer of Io's surface. Volcanic eruptions may spew over 62 miles (100 km) above the surface, since Io has neither substantial atmosphere nor strong gravity. But the surface of this moon is geologically one of the youngest in the solar system. Much of Io close up would look like the painted desert. Much of its coloring is due to sulfur. One scientist, in fact, described it as looking like a pizza. Much of the volcanic material is sulfurous in nature. And probably 220 pounds (100 kg) per second is spewed off into space.

Europa resembles a cracked white billiard ball. It seems to be covered with an ocean of ice. The surface is flat, stretching from

horizon to horizon. Very few craters exist on Europa, making it another moon with a very geologically young surface.

Ganymede seems to have both old and new terrain. Scientists have been fascinated to note tectonic blocks of ice which apparently move over the surface of that moon in much the same way that the continental plates drift on the surface of the Earth.

Callisto is covered with meteor craters. The largest of these is 124 miles (200 km) across. Scientists surmise that a huge meteor struck Callisto perhaps 4 billion years ago. The impact briefly melted some of the ice crust, which rushed in to fill the crater. Other than this, there has been little change on the surface of Callisto over the past four and a half billion years.

The Voyagers have added much to scientific knowledge of the moons of Jupiter as well as the giant planet which they orbit. In encountering Saturn in November 1980, Voyager 1 flew to within 30,000 miles (50,000 km) of the three brightest rings which surround the planet.

Saturn's rings have long puzzled astronomers. It was, therefore, part of the Voyager's mission to explore the rings and confirm the almost-invisible D ring which lies next to the surface of the planet. The rings which range outward from the planet's surface are D, C, B and A. C, B and A are the brightest and most visible rings. But beyond them are the fainter F, G and E rings. Through Voyager's electronic eyes, scientists discovered hundreds of ringlets within the rings. Between the outer B ring and the inner A ring is what looks like a clear space. This space, discovered by a Franco-Italian astronomer in 1676, is known as the Cassini Division (after its discoverer). Scientists had thought the Division was empty space. Voyager showed at least three dozen ringlets within the area.

The Voyagers would also make a careful study of the 15 moons which orbit Jupiter. The largest of these satellites is Titan. Larger than the planet Mercury, Titan is known to have an atmosphere. Some believe that life could have evolved on this moon. Voyager 1 found Titan cloaked in clouds. But other scientific instruments reveal more of Titan's mysteries. Data from the radio-science team plus that from the infrared spectroscopy and radiometry team revealed that Titan's atmosphere is largely nitrogen, like that of the Earth. The atmosphere seemed at least as dense as that of the Earth. And it contained hydrogen cyanide, a necessary building block for the more complex molecules that make up life. Titan's surface, however, is hardly an hospitable environment. The temperature is −292 degrees Farenheit (−144°C), the atmospheric pressure is 4.6 times the Earth's. Still, it presents a frozen record of what Earth's ancient atmosphere was like.

Saturn is much like Jupiter in composition, consisting mainly of hydrogen and helium. In appearance, Saturn is nothing like the brilliantly colored Jupiter. It may be obscured by a thick haze. But it does have a similar 'great red spot' and bands of lighter and darker clouds, and the strong jet streams that sweep around its equator traveling at a thousand miles (1600 km) an hour. Saturn also has seasons, perhaps because it is tilted.

Whatever else the Voyagers have accomplished or will accomplish, they have bridged the gap between known and unknown. The results of their encounters will take years to study and decipher. Yet they have shed a light on the darkness of space and introduced Mankind to the distant planets.

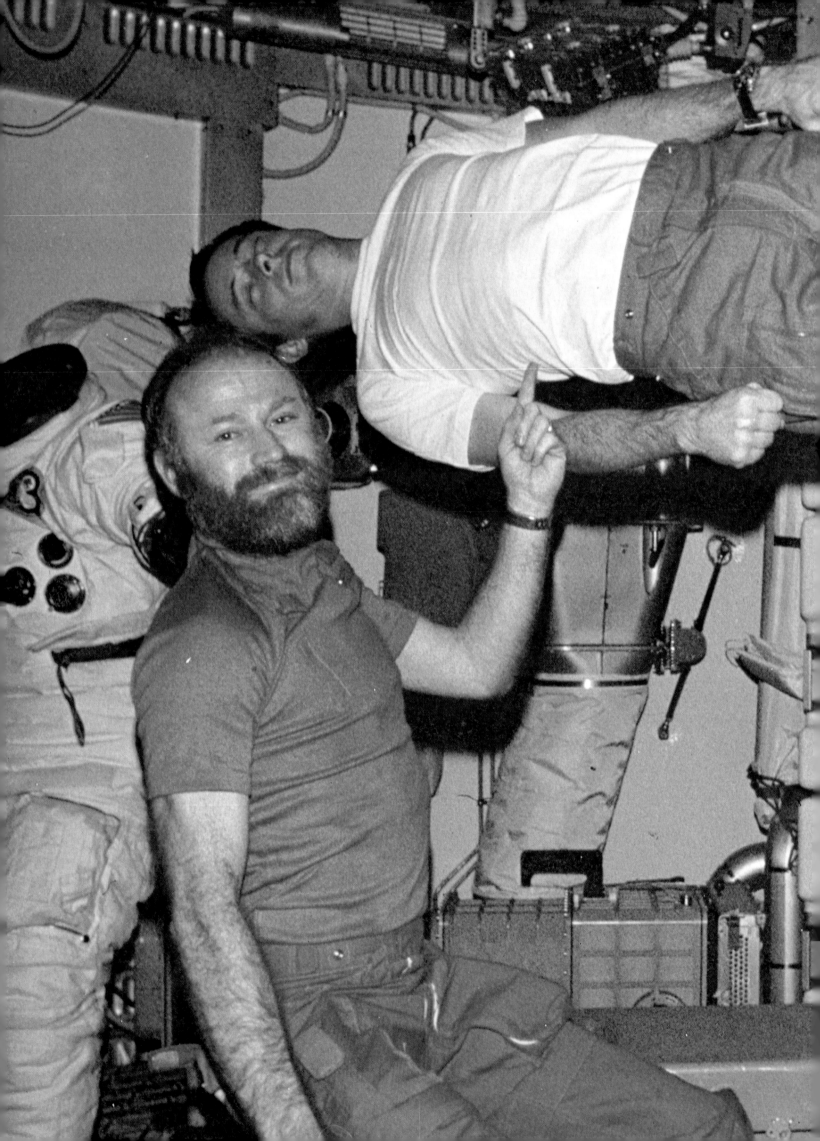

Skylab

With the success of the moon landings, NASA unveiled yet another space program. Called 'Skylab,' the aim of this program was to establish an orbiting space station which would allow astronauts to double the time they were spending in space. It would allow for more complex studies of the effects of space on the human body as well as enable clearer observations of the sun, moon, stars and planets. It was an ambitious program as well as an expensive one. And it nearly fizzled in the first launch.

On the morning of 14 May 1973, a Saturn V rocket stood poised for liftoff on launch pad 39A at the Kennedy Space Center. In an instant, brilliant orange flames burst from five bell-shaped nozzles at the bottom of the rocket. Even at a distance of three miles (4.8 km) the roar was deafening. Fuel pumps working with the strength of 30 diesel locomotives pumped nearly fifteen tons of kerosene and liquid oxygen into the rocket's five engines. As they did so, the thrust began to build. When it reached 7,700,000 pounds (34,500,000 N), the arms which held the rocket would be released. Taller than a 30-story building, the rocket rose slowly, ponderously into the air. Its speed grew as it rose higher into the afternoon sunlight.

The Saturn V was the same type of rocket used in the Apollo Space Program. Unlike the flights to the moon which required a three-stage booster, this rocket contained only two stages full of fuel. The third stage contained Skylab, destined for an orbit 270 miles (435 km) above the Earth's surface. Skylab measured 48 feet (15 m) long and was 22 feet (7 m) in diameter. It was divided into two stories and contained about as much room as a medium-sized two bedroom home. Compared to the Apollo command module, it was downright roomy. It contained living quarters, plus a kitchen and a workshop. A mount for telescopes was attached at one end. There was also a spot where spacecraft would dock with the station and an airlock, a kind of door, that would allow an astronaut to go in and out into space.

Watching what seemed to be a picture-perfect launch were Joseph P Kerwin, a

Right: A two-stage Saturn 1B rocket carrying the Skylab 3 astronauts lifted off at 7:11 am EDT, 28 July 1973.
Previous spread: Astronauts Gerald P Carr (left) and Edward G Gibson (floating) in the forward experiment area in Skylab 4.

144

Above: Skylab 2 Commander Charles Conrad Jr watches a technician check his spacesuit pressure in preparation for his participating in the 'dry' or unfueled portion of the Saturn 1B liftoff.
Right: The Command Module for the Apollo Soyuz Test Project (ASTP) goes through a checkout prior to mating with the Saturn 1B launch vehicle.
Below: The Skylab II EVA near the end of the 28-day mission. Astronauts Charles Conrad Jr and Paul J Weitz went outside, where Conrad retrieved the solar telescope film canisters to stow them aboard the CM for the return to Earth.

Navy flight surgeon, Charles 'Pete' Conrad, a Navy test pilot and astronaut, and Paul J Weitz, a Navy combat pilot and Viet Nam veteran. The three men were jubilant over the launch, for they were scheduled for liftoff on the following day as the first crew of the Skylab Mission.

What was not visible to anyone on the ground, however, was the air pressure building under the thin aluminum sheet which protected Skylab from bombardment by tiny meteorites. As the speed of the rocket increased, the pressure grew. The aluminum, lying like a roof over the top of the station, suddenly peeled up. Caught in a supersonic airflow, it was ripped away. As it did so, it smashed into the 'wings' which contained thousands of solar cells. These would convert solar heat into electricity to power Skylab. The tiedowns which held the right wing were ripped away and the wing opened partially. It was then right in line with the exhaust flames from the second stage, which burned off the damaged wing. A slender strip of metal from the aluminum shield had wrapped around the other wing, leaving it jammed against the station.

No one on the ground was aware of the problems. Conrad, Weitz and Kerwin went back to their preparations for the next day's flight. Meanwhile, Mission Control at Houston was tracking the flight. The engineers there had picked up the first sign of trouble:

Below: Astronaut Paul J Weitz checks out the 'bike' used to check the astronauts' mechanical effectiveness.

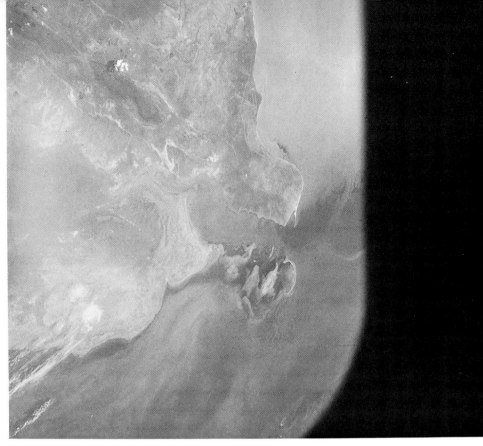

Above: A picture from the Apollo Soyuz Test Project showing the Kazakh SSR, the Caspian Sea, Poluostrov Buzachi and Mangyshlakskiy Bay.
Left: The Skylab Orbital Workshop being prepared for its mating with the SII Stage of its Saturn V Launch Vehicle —4 October 1972.
Below: The launch of the Apollo Soyuz Test Project Saturn 1B Launch Vehicle —15 July 1975

tiny sensors on the shield showed it was vibrating badly. Abruptly, the signals stopped. The engineers guessed that the shield had been torn away. On a more positive note, the space station had achieved orbit exactly as planned.

Within the 56 foot (17 m) nose of the space station was housed a telescope mount, 15 feet (4.6 m) tall, and two modules. Over this 26,000 pound (11,800 kg) nose were fitted four aluminum shells, carefully streamlined for flight and designed to protect the mount. Five minutes after Skylab had attained orbit, these were blown away by a small explosive charge. This exposed the 24,656 pound (11,200 kg) telescope mount. Acting on a signal from Houston, the mount rotated 90 degrees and locked into position on top of the modules in front of Skylab.

Now another signal and four wings measuring 102 feet (31 m) from tip to tip unfolded from the telescope mount. Once locked into position, they resembled the blades of a windmill. These wings contained 164,160 solar cells. Sunlight falling on these cells would be converted into 10,500 watts of electrical power. This would provide about half the power for Skylab. The rest would come from cells on the two wings along the sides of the workshop and living quarters. These, unfortunately, were the two wings which had been damaged when the shield ripped away. Signals from Houston to open the wings met with no response.

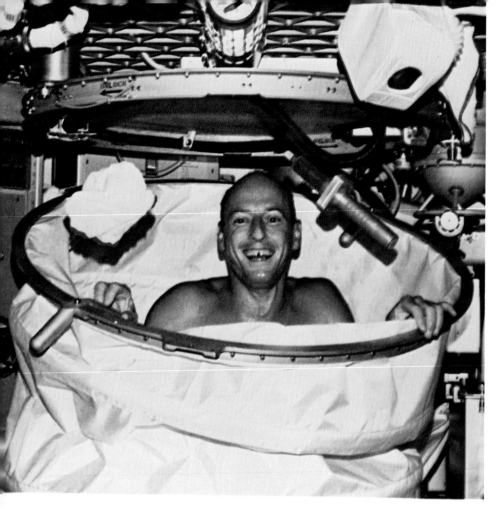

One of the first things that was done was to signal the small thrust rockets to fire and turn the exposed portion away from the sun. By keeping Skylab tilted at a 50 degree angle, the exposed portion remained shielded from the sun, while the solar panels on the telescope mount received enough light to generate power. Temperatures within the ship dropped from an average of 130 degrees Fahrenheit (54 °C) to 105 degrees Fahrenheit (41 °C). This wasn't exactly cool, but it gave the engineers time to come up with a replacement for the lost shield.

One idea was for a giant balloon which would cast a shadow over the space station. It would be carried uninflated aboard the spacecraft with the astronauts. Once they arrived at the station, it could be positioned and inflated. The only problem was that balloons 42 feet (12.8 m) long and 11 feet (3.4 m) wide were not available. Making such a balloon would take too long, so that idea was scrapped.

Dozens of ideas were proposed by NASA engineers, private contractors, and companies who were not even involved with

The space station, now moving at 17,500 miles per hour (28,160 km/h) was soon out of range. The signal was sent again as Skylab passed over Australia. There was still no response. A third signal was sent as the space station completed its first orbit. The wings did not open. Skylab was in trouble.

The wings would not open and without the aluminum shield to act as insulation from the Sun's rays, the temperatures both inside and outside of Skylab began to rise. The engineers at Houston pondered on what to do. The fuel cells on the wings could be replaced by fuel cells and batteries aboard the Apollo spacecraft which would ferry the men to Skylab. The heat was another matter. Temperatures within the space station could easily reach 190 degrees Farenheit (88 °C). The effect of such heat would spoil the food and medicines stored inside. Temperatures on the metal outside might go as high as 325 degrees Farenheit (163 °C), causing buckling and tearing. The heat could cause the styrofoam insulation in the walls to break down and release poisonous gases. Clearly, the number one priority was to find some way to cool the space station quickly. The astronauts' launch was postponed to 20 May and then to 25 May as the NASA engineers tried to come up with an idea to shield the space station from the fierce heat of the sun.

Above: Astronaut Charles Conrad Jr, the Skylab II commander, smiles happily for the camera after a hot bath in the shower facility in the crew quarters of the orbital workshop of the Skylab I/II space station cluster.
Below: Weightless Charles Conrad Jr pedaling the 'bike' with his hands.

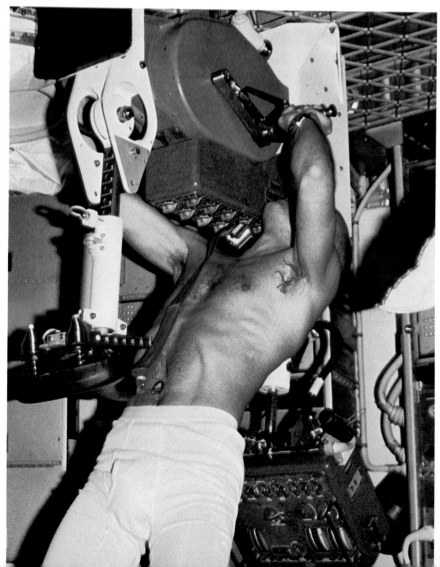

the space industry. Near Houston, workers at the Johnson Space Center came up with a sail-like shield made of aluminized fabric. It could be installed as the astronauts leaned out of the hatch of their Apollo spacecraft. At the Marshall Spaceflight Center near Huntsville, Alabama, engineers came up with another idea. This was a 20×20 foot $(6 \times 6$ m) foil-like shield mounted on an A-shaped pole framework. Yet another idea came from Jack Kinzler, who was chief of the Technical Services Division at Johnson Space Center. Kinzler proposed a 22×24 foot $(6.7 \times 7.3$ m) aluminized nylon parasol. It could be put in place without the astronauts ever getting out of their spacecraft. These three ideas were considered the most workable solutions. All three would be carried into space. One of them would surely be right.

Top NASA officials came up with another bold idea. Weitz, Kerwin and Conrad would blast off from Cape Kennedy on the morning of 25 May. It would take seven hours for them to catch the orbiting space station; once they did, they would maneuver their

Right: Apollo 17 Command Module Pilot Ron Evans demonstrates the squeezable 'glass' that enables the astronauts to drink in zero gravity. Apollo 17 was launched 6 December 1972 and was the last of the Apollo-Saturn missions.
Below: Skylab II backup mission command Russell L Schweickart at the Mission Control console during the Skylab mission —29 June 1973.

spacecraft to the jammed wing. They would then take turns trying to release the wing. The next day, they would install the sunshield. Kinzler's parasol was first choice because of its simplicity. Repairing the Skylab in space would be a difficult and dangerous operation. One astronaut would have to hang out of the open hatch while another held him by the heels as he worked. The third astronaut would have to maneuver the spacecraft near enough but not too near the area of work. If they couldn't repair it, there would be a 20 million dollar hunk of metal uselessly circling the earth.

At 3:30 in the afternoon, 25 May, Pete Conrad sighted the Skylab, 259 miles (417 km) above the Pacific. With care, the 34 foot (10.4 m) long Apollo spacecraft circled the orbiting station. The astronauts could see the meteoroid shield curled around the partly opened solar wing. The other wing was completely gone. 'At first inspection, it looks like we can fix the solar wing,' Conrad reported. 'I think we can take care of it with the Stand-up EVA.' In the parlance of space, that meant extravehicular activity while standing in the open hatch of the Apollo spacecraft.

At the front of the space station, Conrad carefully brought the Apollo nose to nose with Skylab. A probe on the spacecraft was fitted to a socket on the station. Twelve powerful latches on the spacecraft locked onto a ring on the station. The Apollo was now firmly docked to Skylab. After a period of relaxation and dinner, it was time to tackle the broken wing. They would all have to wear the bulky spacesuits, helmets and gloves, because once the hatch was opened on the Apollo, all the air would be gone. There would be only the vacuum of space. With the spacecraft now undocked, the three eased up to the crippled wing. Weitz stood in the hatch, head and shoulders clearing the doorway. Below, Kerwin firmly held his ankles. With a special long-handled tool, Weitz attempted to pry away the metal wrapped around the wing. But pull and prod as hard as he would, the metal strip refused to budge. In his attempt, the two ships nearly collided. Conrad hastening to avert trouble, started to pull away and Weitz found himself in the very hairy predicament of being stretched between Skylab and Apollo. Conrad quickly corrected his course, and a rather shaken Paul Weitz climbed back down the hatch. Repair of the damaged wing was impossible. They were flying into sunset and it would be too dark

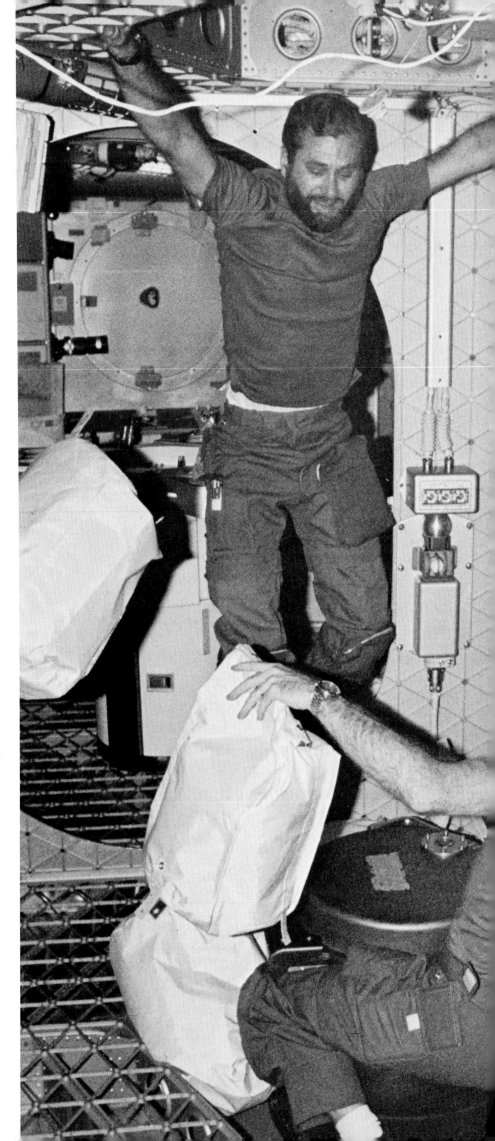

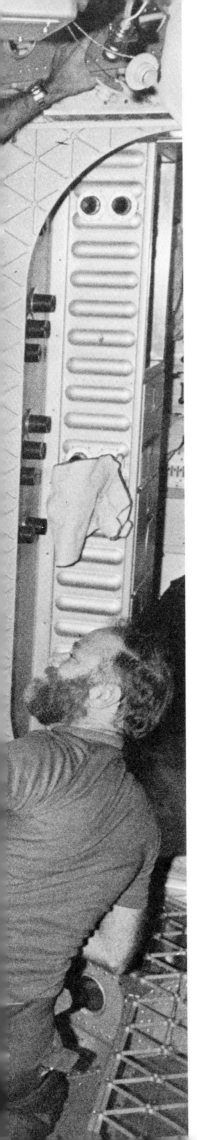

to see. They returned to dock, but found it impossible to get the two craft to lock together again. After several tries, Conrad decided to use the small thruster rockets to force the craft together. This resulted in a secure latch. Linked together, the spacecraft now orbited the Earth once every hour and a half.

While the men aboard the Apollo slept, the ground team signaled pumps aboard the space station to begin building up the oxygen and nitrogen pressures inside. The atmosphere aboard Skylab, unlike that on Earth, was 30 percent nitrogen and 70 percent oxygen. (Earth's atmosphere is 78 percent nitrogen and 21 percent oxygen.)

The next morning the astronauts repaired the docking system. Once this was accomplished, they explored their new home thoroughly. The first module which they entered was called the multiple docking adapter or MDA. In it were two docking spaces, one for the Apollo spacecraft and one for a rescue craft if it were ever needed. The room was 17 feet (5.2 m) long and 10 feet (3 m) in diameter. Within it were the controls for telescopes and various instruments used for studying the Earth. Other equipment included a special furnace which would be used with metal working, welding and growing crystals.

At the far end of the MDA was the airlock module (AM). Housed in this 18-foot (5.5 m) room was all the equipment for sending data and messages back to Earth. There were controls for the electric power. There were systems for regulating the temperature and ventilation. There was also a short tunnel with doors at either end, the airlock. Within the airlock was a door leading to the outside. When any of the men was scheduled for EVA, he left and reentered through this door. There were numerous jobs which required working on the exterior of the station.

The door at the far end of the airlock led to the orbital workshop (OWS). This was a large compartment, 48 feet (14.6 m) long and 22 feet (6.7 m) in diameter. Because of its size, it was divided into two floors. The work area was on the upper floor, while the lower floor contained the kitchen and sleeping quarters.

Left: Astronaut William R Pogue of Skylab 4 holds onto the OWS crew quarter ceiling as he prepares to jump on the airlock hatch cover to force a trash bag farther down into the air lock. Gerald P Carr is assisting. A third trash bag is floating in the zero-gravity environment near Pogue's right leg— 12 February 1974.

Once the space station was checked out, the next job was to install the sunshade. It was an easy enough matter getting the parachute's container through a special airlock opening in the wall of the workshop. Then using the handle or main shaft of the parasol, Conrad and Weitz pushed the sunshade through the outer wall of the OWS. They added rod extensions as needed. Once the four telescoping ribs of the sunshade cleared the outer wall, powerful springs snapped them into their full extension, a 22 × 24 foot (6.7 × 7.3 m) rectangle. There were a few wrinkles on one side. Conrad and Weitz tried to shake them out, moving the parasol up and down. Finally they left the parasol stretched out against the surface of the workshop. The heat of the sun pressed out the wrinkles. At last everything aboard Skylab was in good shape. The effect of the parasol was almost immediately noticeable. In the first hour and a half, temperatures on the outside skin of the station dropped 50 to 60 degrees Fahrenheit (10–17°C). The astronauts were now cleared to complete their 28 day mission. A 2.6 billion dollar project had been saved by a 75 thousand dollar parasol. Now that they knew the Mission was secure, the astronauts set about house keeping. All of the lights, aircleaners, air conditioners, water, toilets and scientific equipment were turned on and checked to make sure they functioned properly. Because any loose item would fall or fly around during the launch and booster flight, NASA technicians had packed everything into boxes and tied the boxes to the floors of the station. All of these things now had to be unpacked and stored away. There were at least 20,000 different items ranging from 2000 pounds (909 kg) of food to 1200 aspirin. The one pleasant surprise in all this was that the heavy boxes were so easy to handle at 0 gravity. On the other hand, when a plastic bag with corn in it broke, the kernels went flying everywhere. Cleaning up a messy spill at 0 gravity was no fun.

After moving in, there were various medical tests to be performed. The human heart, which on Earth works against the pull of gravity, tends to get out of condition in space where it doesn't work against this force. The muscles, also, will deteriorate in 0 g conditions. The astronauts worked out on an exercise bicycle to keep their muscles in condition. They also took some blood samples to determine the effects of long term weightlessness on blood and disease resistance.

US Navy personnel hoist the Command Module containing Skylab 4 astronauts aboard the prime recovery ship USS *New Orleans*—8 February 1974.

which, because of the damaged wing, was in short supply. The energy crisis was beginning to take a serious toll on the Mission.

The flight directors back in Houston began to plan another attempt to fix the stations broken wing. Twenty-five feet (7.6 m) from the exit door in the airlock module was the metal strap which had riveted around the wing. Would it be possible for an astronaut to walk over and remove that piece of metal? How would he hold himself in place while he worked? What kind of tools would he need? As usual, the engineers on the ground decided to reconstruct the situation as it was in space. To make up for the weightless conditions the astronauts were experiencing, they used a huge water tank at the Marshall Space Flight Center in Alabama. An aluminum strap was fastened to a solar wing on a full-sized model of Skylab. By experimenting with different tools and different techniques, they finally found what seemed the best way to free the crippled wing.

On 7 June the astronauts Conrad and Kerwin climbed out of the airlock in what they billed the 'high wire act of the century.' Using a 25 foot (7.6 m) pole with a pair of metal cutters on the end, they set to work. The action at this point was complicated by loose wires, the inability to maintain a foothold, and getting the cutter's jaws to grip the strap from that distance. Then there was the further confusion of untangling their umbilicals. With the pole at last fastened down at both ends, Conrad made his way hand-over-hand to the broken wing. He checked the position of the jaws on the cutting tool. Then he signaled Kerwin to pull the rope which would operate the cutter. Kerwin, who was hanging by his toes over the edge of the workshop, made the cut. The useless metal strap fell away. Now the wing would have to be broken free, for oil in its hinge had frozen fast in the cold of space. Conrad accomplished this using a rope attached to the telescope mount which he also attached to the wing. One good sharp tug and the wing broke loose, freed at last. By evening, the once crippled wing was beginning to generate power. NASA and the crew aboard Skylab were elated. A major repair had been accomplished. There was one more repair that needed to be made. This was the case of the stuck regulator. The regulator had been designed to recharge a battery. It was supposed to feed current from the solar panel on the telescope mount but

Above: Soviet and American Apollo-Soyuz Test Project directors visiting the spacecraft launch vehicle and testing facility at the Baikonur (USSR) launch site on 19 May 1975.

Left: An artist's concept of the Apollo-Soyuz docking. The purpose of the exercise was to test the jointly designed international docking mechanism and to gain experience for the conduct of potential joint flights by US and USSR spacecraft, including, if necessary, rendering aid in emergency situations.

Overleaf: An interior view of the Docking Module Trainer for the ASTP at the Johnson Space Center in Houston. Astronaut Donald K Slayton (right) and Cosmonaut Valeri N Kubasov.

Skylab, as its name implied, was an orbiting space laboratory. In addition to determining the effects of weightlessness and space travel on the human body, there were celestial and solar observations to be made. The human eye cannot look directly at the Sun, but the telescope can. In this place, high above the clouds, haze and fog of the Earth's atmosphere, scientists could get a first clear view of the Sun. Eight different telescopes and instruments were focused on the sun from Skylab. Moreover, the instruments could be focused on the Earth as well as the Sun and stars. It would be possible to do a survey of the Earth from space. Using the Earth Resources Experiment Package (EREP), the astronauts could gather information vital for mapping, weather forecasting, crop surveying, detecting crop and tree diseases, forests, mineral deposits and many other things vital to conserving Earth's resources. The surveys were very successful, but the instruments needed a great deal of electrical power,

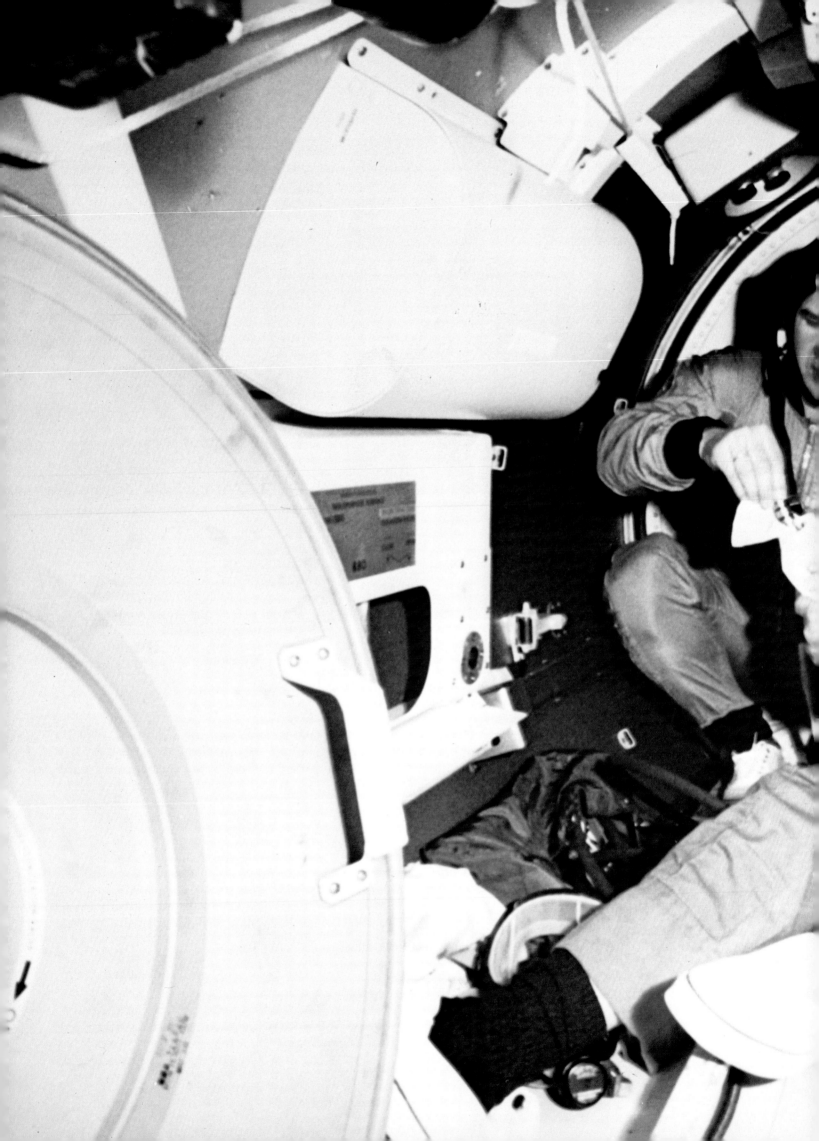

NASA engineers wanted to know how the orange aluminized material of the sunshade was holding up. Conrad was supposed to attach the stuff to one of the struts on the telescope mount. The little sample would be returned to Earth for examination by the next crew of men. The fabric proved difficult to attach. In working to get it fastened down, Conrad's heartbeat shot to 150 beats a minute. This is more than twice the normal rate of 70 beats per minute. Everything calmed down, however, once the strip of cloth finally was in place.

Despite the mishaps, all had really gone quite well with this first Skylab Mission. Even a jammed-up trash lock in the last days presented no long-term problem. After 28 days in space, they all noticed light-headedness and a heavy feeling, a normal reaction after being weightless for so long. They had suffered some calcium loss, though not enough to affect the strength of their bones. Their heartbeats were faster than normal but these soon slowed. All had lost weight but no one more than eight pounds (3.6 kg).

There were two other Skylab Missions. The second one, which left the Earth on 28 July 1973, was plagued with problems. All three of the astronauts, Alan Bean, Jack Lousma and Owen Garriott, were very ill with motion sickness early on in the mission. Then the service module of their spacecraft began to leak nitrogen tetroxide. This combined with hydrazine formed the fuel for the thrusters. For a time, NASA planned a rescue mission, but as time continued, the rescue mission was scrapped. After 59 days, 11 hours and 9 minutes and numerous scientific accomplishments, the men returned to each on 25 September 1973.

On Friday 16 November the last Skylab crew was blasted into space. It consisted of Jerry Carr, Ed Gibson, and Bill Pogue. Their launch had been scheduled for earlier but cracks around the bolts which held the fins to the rocket body forced a cancellation. They would spend 84 days in space.

On the whole, Skylab had been an exciting adventure. Much had been learned about human endurance and inventiveness. The scientific instruments and cameras had functioned well. The surveys of the Earth had produced much new information which could only have been gained from space. Pictures of magnetic storms on the Sun gave scientists another view of Earth's nearest star. The huge flares and sunspots which occur in regular cycles had long been of

Above: A camera located at the mobile launcher's 360-foot level recorded this view of the Saturn 1B rocket carrying the Skylab 2 astronauts on their first leg of the journey to the orbiting Skylab space station.
Opposite: The two crews of the joint US-USSR ASTP docking in Earth orbit mission made the mission an unqualified success. Kubasov (top) and Leonov in the Orbital Module.

it had not worked since the mission began. Finally, as a what-have-we-got-to-lose measure, Pete Conrad went after the regulator with a ballpeen hammer. Amazingly, after a few whacks, the regulator began to function.

Six of the eight cameras now needed to be loaded with film. The smallest film magazine was nearly the size of a thick encyclopedia. The largest was about like an over-night bag and weighed 60 pounds (27.3 kg). The film-changing was not difficult. In fact it was rather fun as long as the men remembered to move slowly.

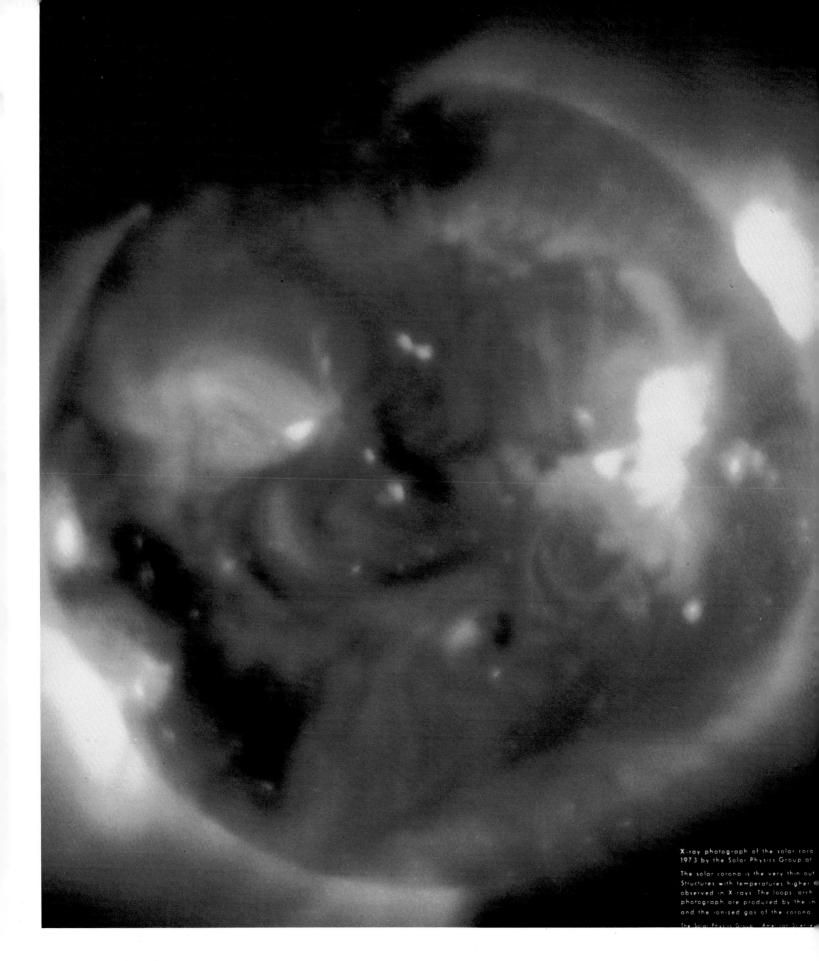

X-ray photograph of the solar coro
1973 by the Solar Physics Group of

The solar corona is the very thin out
Structures with temperatures higher
observed in X-rays. The loops arch
photograph are produced by the in
and the ionized gas of the corona

Above: A view from Skylab of the X-ray corona of the Sun.
Opposite: The ASTP Apollo's two main parachutes collapse as the spacecraft touches down in the Pacific Ocean.

interest. There were also the investigations and experiments which high school students had suggested. As Skylab director Bill Schneider said, 'The real payoff—the reason for the whole Skylab project—is the data we've obtained.'

On 12 July 1979 Skylab returned to Earth. It rose as a brilliant comet in the southwestern sky and fell in a shower of fiery scraps over central Australia and the southern Australian coast. Phase One of the space program had come to an end.

Space Shuttle

Launch pad 39 basked in the brilliant glow of powerful searchlights. Within the circle of light stood a 154 foot (47 m) long tank, measuring 28.6 feet (8.7 m) in diameter. The tank was flanked by two rockets which measured 150 feet (45.7 m) long by 14 feet (4.3 m) in diameter. Mounted on the tank, its nose pointed skyward, was an aircraft which looked like a DC-9. The whole thing glowed with an unearthly whiteness. Out of the Florda darkness a voice over a loudspeaker droned, 'T minus four hours and counting.'

Three miles (4.8 km) distant from the launch pad at the Kennedy Space Center, 35 hundred news reporters from the US and more than 70 nations waited for liftoff. The space shuttle Enterprise was now in the final hours of countdown. Within an hour, the astronauts would board the ship. They would be carried by elevator to the cabin of the shuttle orbiter. They would be strapped into specially contoured seats and then, 275 feet (83.8 m) above the ground, they would begin running down a long checklist. A web of gauges and dials stretched before them, each of which monitored a special function of the craft. The astronauts began with the guidance-alignment checks. There were myriad details to be checked. The time they had alotted (180 minutes) was none too long. As the countdown reached the final minutes, the tension mounted. Finally the test supervisor's words came over the loudspeaker. 'Cleared for launch.' The final two minutes of the countdown would be handled by computers. The propellant tank vents were closed. Within their cabin, the astronauts glanced across the instrument panels. Every indicator was green. The time was now T minus 15 seconds. A thousand cameras, set to record the launch, began to whirr. The countdown proceeded. There was a sudden roar, a flash of flame, flaring brilliant orange. A thundering shock wave echoed off the Cape.

The 7,000,000 pound (3,182,000 kg) shuttle craft rose heavenward much more rapidly than did the old Saturn V rockets. Unlike the Saturn rockets, the strap-on rockets which would lift the Enterprise into orbit used solid fuel. These solid fuel rockets would attain their full thrust, 5.3 million pounds (23,574,400 N), in .04 of a second.

Previous spread: The space shuttle *Columbia* is silhouetted against the morning sky as the space-craft prepares to touch down on Runway 22 at Edwards Air Force Base 5 January 1983.

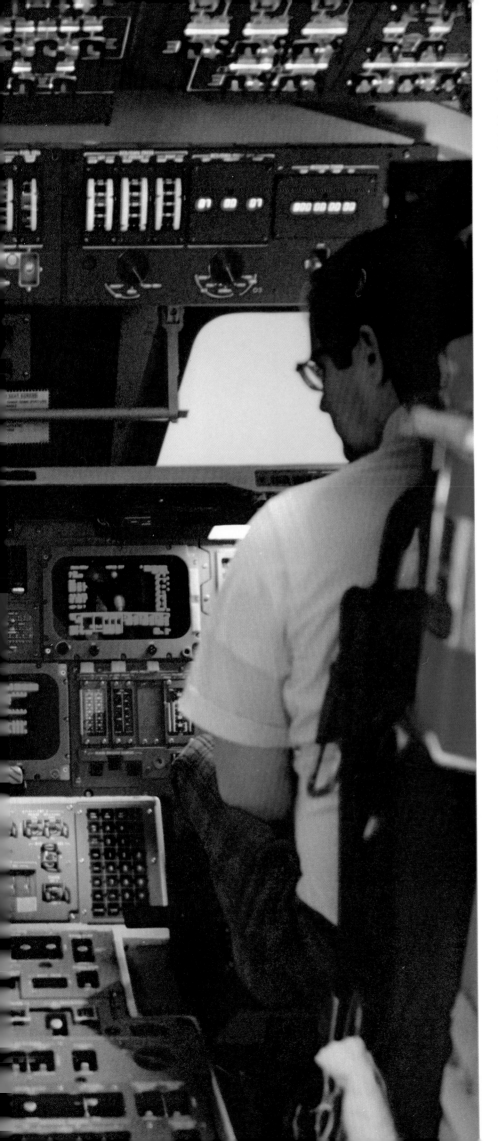

After two minutes of flight, when the shuttle was 27 miles (43 km) above the Earth, the two booster rockets burned out. At that point tiny explosive devices fired and separated them from the tank. As they fell back toward the sea, parachutes deployed to slow their descent and mark their location. A recovery ship moved swiftly to pick them up and tow them back to shore where they would be reconditioned and used again.

The main engines would burn for another six minutes, carrying Enterprise to the edge of space. Once they shut down, the huge tank would be jettisoned, to tumble in a 10,000 mile (16,090 km) arc toward the Indian Ocean where it would break up before entering the Earth's atmosphere. Two small orbital maneuvering engines would then push the shuttle craft into orbiting speed.

The shuttle program had actually begun in the late 1960s. Overshadowed by the lunar landings, the shuttle program remained under wraps until the early 70s. The moon landings were beginning to lose their luster. The American public, beset with the energy crisis, was beginning to ask hard questions. With the poverty, hunger and pollution problems on Earth, was the space program really worth the billions of dollars spent on it? Could the exploration of space solve any of the more immediate problems on Earth? It was difficult to justify the expense of the space program in the face of all the other difficulties the nation faced. But it was under these circumstances that the second stage of the space program developed in the form of the shuttle.

As it turned out, the shuttle was complex, costly and subject to problems just as the earlier space programs had been. What was true of the shuttle that was never true of any of the earlier spacecraft was that it could be used over and over again. An excellent selling point, when one considers the cost of the space shuttle was 9 billion dollars!

The shuttle is actually a kind of cargo ship which can carry satellites, military hardware and passengers into space and bring them back. It can replace other satellite launching vehicles, retrieve malfunctioning satellites for servicing, and carry

Left: An interior view of the Space Shuttle mission simulator in the mission simulation and training facility at the Johnson Space Center in Houston. NASA personnel are seated in the pilots' positions.

165

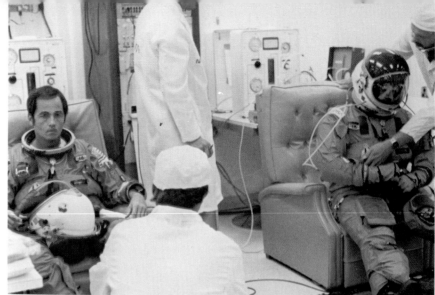

MAXIMUM CAPACITY 4 PERSONS OR 1000 LBS.

Top left: Astronauts John W Young (foreground) and Robert L Crippen train for their 1981 *Columbia* flight.
Top right: A technician adjusts Crippen's space suit legs while Young's suit is connected to the life-support system lines.
Above: Crippen and Young take part in a simulated rescue operation.
Far left: Astronaut John W Young, commander of the STS-1 *Columbia* space shuttle.
Left: Robert L Crippen, the STS-1 pilot, leaves the Space Shuttle Orbiter 102 *Columbia* 14 April 1981 following a flawless landing on Rogers Dry Lake at Edwards Air Force Base. John W Young stands at the foot of the steps to greet his crewmate.

Above: An overall scene of the celebration following STS-4's successful landing on 4 July 1982.
Left: Soviet Communist Party leader Leonid I Brezhnev and President Richard M Nixon, during ceremonies at San Clemente, examine plaques presented by Skylab astronauts Charles Conrad, Paul Weitz and Joseph Kerwin (left to right).
Below: The five STS-7 crew members on their return 24 June 1983. In the blue jumpsuits, left to right, are Norman E Thagard, John M Fabian, Frederick H Hauck, Sally K Ride and Robert L Crippen.

Above: High above the Kennedy Space Center, *Challenger* heads toward its second mission.
Right: *Challenger* lands after its second mission.

a variety of equipment into orbit. Many of the advanced program planners at NASA see the shuttle as the beginning of a space program for the general public. 'I'm convinced that by 1990 people will be going on the shuttle routinely—as on an airplane,' says Robert Freitag of NASA. There is also talk of establishing industry in space. Certain kinds of techniques, such as welding, performed in the weightless conditions of space, produce superior results. Astronauts in space shuttles can deploy and service the complex switching stations that television and telephones will use. In the not too distant future, huge telecommunications platforms will be constructed in space. These will be able to handle 250,000 phone calls simultaneously. They would also allow Earthbound viewers to tune in any television program in the world. The plans for these platforms are already underway and the space shuttle will play a large part in them. There is also the possibility of using

the shuttle in deploying military satellites. Whatever the future of the shuttle craft, much will depend on NASA's budget and the interest of private industry.

Basically, the shuttle orbiter is the largest craft ever put into orbit. It posed no small problem for designers. It would have to be launched like a rocket, orbit like a spacecraft and return to Earth flying much like an airplane. At nine minutes before liftoff, the shuttle's quad-redundant bank of computers take over the flight. 'Quad-redundant' means that there are four computers. There is also a fifth as a back-up. They operate until just a few moments before touchdown. At some of the most critical points during the flight they will process as much as 325,000 operations per second. Each of the four computers processes the same information. They must all agree. If, however, one disagrees, it will be shut down. If one of the other three also disagrees, the majority still carries. Should the two remaining compu-

ters also disagree, the fifth will be brought in to make the decision. The use of such a computer system has eliminated the need for the dozens of mission controllers who monitored the Apollo flights. Four ground controllers can actually handle the flight of a shuttle craft. Within the craft, three television screens report data on the trajectory, guidance control, electrical, environmental and hydraulic systems. Should a problem arise, a warning beep will sound in the astronaut's headset. A master alarm will flash. Lights on the control panel will show the source of the problem. Using a computer keyboard, the astronaut can then call up all the data regarding the problem. Orbiters are too complex to monitor constantly. The computers allow the astronauts to work on more immediate tasks.

Left: A view of the open cargo bay of the orbiter *Challenger* in the Orbiter Processing Facility.
Bottom left: The STS-1 being moved out of the Vehicle Assembly Building, 29 December 1980.
Below: Crippen's wife greets him after the landing.

Above: Aboard the STS-7 —20 June 1983. Left to right: Frederick H Hauck, Norman E Thagard, Sally K Ride.

Opposite: The orbiter *Challenger* is shown with a special lifting sling attached as it is raised vertically inside the Vehicle Building at the Kennedy Space Center in preparation for mating with the external tank and solid rocket boosters.

Except for the TV screens, the 1400 switches and circuit breakers, the cockpit of an orbiter looks almost like that of any transport plane. Behind the cockpit, however, is another small area filled with more control boards and a view out over the long cargo bay. This is the office of the mission specialists like Sally Ride—people who have a particular job to accomplish on a particular flight. From their window view, they can watch the remote manipulator arm as they raise it to lift another satellite into orbit.

There are two levels within the orbiter cabin, both quite compact. Beneath the flight deck are the living quarters. They include three seats, a galley, a washroom, sleeping quarters and an airlock which exits into the payload bay. The decor could be described in one word—stark.

The exterior of the shuttle craft, however, has been a technological breakthrough. Because it reenters the Earth's atmosphere at high speeds and high temperatures (sometimes higher than 2500 degrees Farenheit or 1370°C), there was a need for some type of heat shield. Obviously, the old type used on the earlier spacecraft wouldn't do. Those chemical shields absorbed heat, charred and flaked off. The orbiter would need something more durable. It would be reentering the atmosphere as many as 100 times. The solution was to use tiles made from fibers of 99.7 percent pure silica glass. Fibers of that purity conduct almost no heat. Heat on the surface of these tiles escapes into the air. Heat within remains trapped for hours amid the silica fibers. Thirty-four thousand such tiles, covered with a heavy-duty reflective

Above: A photograph taken from the Earth-orbiting *Challenger*. Shown are parts of Australia. Opposite: The right-hand Maneuvering System pod is shown being installed on Space Shuttle orbiter *Columbia*. The engines in both left and right-hand pods provide the thrust to get the orbiter into final orbit after the main engine cutoff and the jettisoning of the external tank. They also aid in maneuvering. Previous spread: Camouflaged by tiers of work platforms, the Space Shuttle orbiter *Columbia* undergoes systems installation and checkout.

sealer, cover the underside and parts of the orbiter that will be exposed to the severe heat of reentry. Each tile has a varying shape and thickness. They must curve to fit the lines of the shuttle craft. The tiles are tailored and cut by computer. But each one is glued by hand to the orbiter's aluminum skin. It is something like putting together a giant jigsaw puzzle. The glue is crucial. In the past, tiles have fallen off, leaving areas of the shuttle vulnerable to the intense temperature changes.

The tiles were a minor problem to designers, compared to the three main engines on the shuttle. These engines which take over after the solid fuel rockets are jettisoned release as much energy as 23 Hoover Dams. But they are surprisingly light in weight, weighing only about 7000 pounds (3182

kg) each. Without the nozzles they are about 5 feet (1.5 m) high. Over a thousand tiny tubes run up and down the length of the nozzles. During the heat of combustion, the super-cold hydrogen fuel is syphoned off by these tubes and used to cool the nozzle. (Without this cooling, the nozzle would soon melt.) Other tubes then collect the hydrogen and feed it back to the two turbines. The pressure within the engine builds to nearly 7000 pounds per square inch (4830 N/cm^2). Such pressure is necessary to get the maximum thrust from the hydrogen fuel. But it also puts enormous demands on the technology that develops such engines. In test firing after test firing, turbine blades have cracked, rotating parts and bearings have broken down, and sometimes combustion has begun prematurely. The engineers at

175

Rocketdyne, where the engine was developed, needed an additional year just to work out the problems.

There were the aforementioned problems with the tiles, which at one point had to be removed and rebonded to the craft. There were difficulties in developing the computer programs, used in the flight simulators. It was almost as though technology was, for once, racing to catch up with the dreams of the nation.

The one component, the backbone of the whole system, was the huge external fuel tank. Developed by Martin Marietta, it caused no delay whatever in the shuttle's proposed launch date. It is described by some NASA engineers as a 'big tin can,' but that is an over-simplified description. The orbiter and the solid-fuel rockets are bolted to the tank. At launch, the thrust of five engines comes to bear on the inner struts and beams of the tank. It holds 20 times its own weight in liquid hydrogen and oxygen. Its shell must be extremely light yet able to withstand the force of an enormous explosion. The fuel within must be kept at hundreds of degrees below zero to maintain its liquid state. More than a half mile of welds are needed to put the external tank together. After it is assembled, it is sprayed with a foam insulation. This is done because the odd shape of the shuttle creates drag during its liftoff and ascent. Drag creates friction

and sections of the tank can get quite hot.
At temperatures above 350 degrees Farenheit (177°C), aluminum loses strength and
may fall apart. Hence, the foam insulation.

The orbiter looks like a stubby little
airplane. None of the grace of the jet is
apparent at first glimpse. In orbit, it comes
into its own. The launch is rougher than
previous manned space flights due to the
thrust of the solid fuel rockets. The speed of
the orbiter will build as the fuel tank
lightens. The astronauts must throttle the
engines so that the shuttle does not exceed
the speeds for which its wing structure is
designed. As it climbs toward orbit, the
orbiter is flying upside down, on the underside of the fuel tank. This allows for an easier

separation from the tank, using a minimum
of thrust and relatively little G-discomfort.
Once the main engines shut off, there is a
16 second delay before the external tank
separates from the orbiter. Then for one
minute and 45 seconds the orbiter drifts,
waiting for the tank to move safely away.
The astronauts now check the position, the
trajectory and the systems data to be certain
the shuttle will reach orbit. Should it be
necessary, they can abort the flight, fly once
round the earth and land at a desert site in
New Mexico.

Once it is decided to continue the flight,
the auto pilot is given the go-ahead. The
Orbital Maneuvering Subsystems (OMS)
engines take over. The first burn will lift the

shuttle into orbit. A second, 35 minutes later, will change its path to a circular one. Once the shuttle is in orbit, the doors of the cargo bay will be thrown open. A considerable amount of heat is generated by all the electronics equipment. With the doors open, the heat is dissipated into space. Though the doors, made of graphite epoxy, are built to withstand temperature variances from −170 to 135 degrees Farenheit (−76 to 57°C), there is still some possibility they could warp or their latches or motors could fail. Any problem of that kind has to be repaired by an astronaut working outside the ship.

When the shuttle prepares to return to Earth, small reaction control jets turn it around, so that the engines now are forward. As the orbiter passes over the Indian Ocean, the OMS engines make a two minute burn to slow the speed which at this point is about 25 times the speed of sound. The shuttle is now dropping, nose up, with the tile-covered underside poised to absorb the reentry heat. Thirty-five minutes later over Midway Island, the orbiter collides with the Earth's atmosphere. To control its speed, it banks and flies in broad traverses. The strain

is critical. The nose is beginning to drop. By the time the first eager onlookers catch sight of it, the shuttle will be into a steep dive, coming into a landing at an approach ten times steeper than a jetliner. At 78,000 feet (23,800 m), the astronauts switch from automatic pilot to manual control. The landing has to be exactly right the first time. There is no chance for a second try. A shuttle lands just like a glider, with no engines.

The construction of the first orbiter was begun in 1974. It was simply designated Orbiter 101. Its final assembly was under the guidance of Rockwell International, one of NASA's prime contractors. The mid-fuselage section and the cargo bay were built by General Dynamics of San Diego, California. Grumman Aerospace Corporation of Bethpage, New York, constructed the wings. Fairchild Republic of Farmingdale, New York, was responsible for the tail assembly. The three main engines were produced by Rocketdyne, a division of Rockwell International in Canoga Park, California. The final assembly began in August 1975. In September 1976 the public got a first look at the orbiter. On 15 February 1977

Opposite top: The modified X-15 aircraft with new external fuel tanks is rolled out of the North American Aviation factory. The tanks increased the engine burning time to 145 seconds at full throttle, which would raise the speed by some 1300 miles per hour. Opposite bottom: The Columbia returns to the Shuttle Landing Facility at the Kennedy Space Center atop the shuttle carrier, a modified Boeing 747, after a two-day cross-country flight from the Dryden Flight Research Facility following its fifth mission in space. The photo was taken 22 November 1982. Below: A group of X-15 pilots.

the stubby-looking orbiter was mounted piggyback on a 747 jet plane for a series of tests. The cumbersome combination weighed in at a total of 544,000 pounds (247,300 kg). On 18 February the 747 took off with its 'passenger' on a two-hour test flight. The results were actually better than officials had expected. Tests continued for most of the remaining year. The first few were aimed at checking out the aerodynamics of the orbiter. The next series was slated to test the systems. And finally, a series of free flights when the orbiter would be separated from the 747 to glide into a landing on its own. The astronauts for the heart-stopping first test were Fred Haise and Gordon Fullerton. Within an hour after take-off, they were separated from the 747 and completely on their own. They were at an altitude of 4 miles (6.4 km) above the Earth. After a series of careful maneuvers, Haise and Fullerton found they could increase and decrease the speed of the orbiter merely by moving its nose up and down. Then there were two long 90 degree turns to the left to align themselves with the runway below. As easily as a bird born to fly, the orbiter touched down at Edwards Air Force Base, and braked to a smooth stop. In the next two and a half months, four more free

Previous spread: The orbiter
Columbia riding piggyback on
NASA's Boeing 747. The T-38
chase plane at the far left was
flown by Donald 'Deke'
Slayton.

Below: *Columbia* STS-1
crewmembers flight
commander John Young *(left)*
and pilot Robert Crippen *(left)*
have a pre-dawn breakfast in
the crew quarters of Kennedy
Space Center's Operations and
Checkout Building before
suiting up for the historic
first Space Shuttle orbiter
flight on 12 April 1981.

Right: Riding upwards on
seven million pounds of
thrust: the largest solid fuel
rockets ever flown (and the
first ever designed for re-use),
each delivering a peak thrust
of approximately 2.65 million
lbs; and the orbiter's three
liquid fuel main engines, each
delivering 375–470 thousand
lbs of thrust, propel
Challenger on its first
satellite rescue mission.

flights followed. They were each so success-ful, that no more tests were scheduled.

The following spring, 1978, Orbiter 101 was flown across country aboard its 747 mother ship to Huntsville, Alabama. There it received vertical ground-vibration tests. There it was fitted with the external fuel tank and booster rockets and stood on its tail within the test facility. Such a test is designed to simulate the stress and vibration the vehicle will meet during launch. The information gained from such tests allows engineers to analyze flight control stability and the kinds of stresses the shuttle will be facing. Orbiter 101 did extremely well in the tests. The final phase of testing would be a series of six manned test flights into orbit and return to Earth. The testing program would last into 1980. Once the test period was over, the shuttle could, they hoped, begin regular operations.

The Columbia, which has become so spectacularly successful in all recent shuttle flights, has in its lineage, the X-1 and the X-15 of the early 1960s. It was the X-1, a purely experimental aircraft which first

broke the sound barrier on 14 October 1947 with Charles Yeager at the controls. It was the X-15 which would first take men to the lower reaches of space. Built to fly 50 miles (80 km) above the Earth's surface, return through the atmosphere and land at Edwards Air Force Base, the X-15 was the forerunner of Columbia. All of this was going on in the late 1950s and early 1960s, just as the space race was gaining momentum. By the time the Soviets had launched the first Sputnik, there were already plans on the drawing board for the X-20, a ship which would be launched by a gigantic Titan III rocket to orbit the Earth and land at Edwards AFB. The Titan III rocket, however, was a problem. It was several years from full development. It seemed urgent that the United States enter the space race as soon as possible. So a smaller, but fully developed rocket, the Redstone, was chosen to launch, not an aircraft but a space capsule.

In 1969, with the success of the Apollo project, NASA returned to the X-20 project, which became the space shuttle.

Left: *Columbia* lifts off from launch pad 39A at the Kennedy Space Center in Cape Canaveral, Florida. Commander Ken Mattingly and pilot Henry Hartsfield, Jr flew this seven-day mission, which was the final developmental test flight for the Space Transportation System, and also the first orbiter landing on a paved landing strip (previous flights touched down on the Edwards AFB dry lake bed). Several orbiters have been used in the STS. In the order of their actual shuttle program employment, they are *Enterprise, Columbia, Challenger, Discovery* and *Atlantis*.

Enterprise (named for the 'Federation Starship' of the *Star Trek* television series), which was purely an atmospheric glide tester, was never intended for and never used in space flight. *Columbia*, the first shuttle in orbit, was a veteran of six space flights through early 1986 when the program was suspended. *Challenger* was not originally intended for space flight, but it was later converted. A veteran of nine successful missions, *Challenger* was destroyed in the 28 January 1986 tragedy, due to an 'O-ring' sealing failure in one of its solid fuel engines 73 seconds after liftoff. Both craft and crew—Jarvis, McAuliffe, McNair, Onizuka, Resnik, Scobee and Smith—were lost. *Discovery* marked a change in orbiter configuration: many of the heat-shielding white ceramic tiles on this craft were replaced with silicon fiber blankets for a weight savings of one-half ton. *Atlantis* is the newest orbiter.

Above: Photographed at the Johnson Space Center's shuttle mockup and integration laboratory, first American woman in space Dr Sally K Ride is shown during a pause in training for her historic flight. Physicist Ride was a mission specialist on the *Challenger* STS-7, launched on 18 June 1983. Besides Dr Ride's accomplishment, STS-7 achieved several other 'firsts': the first use of a remote manipulator arm to launch a satellite (a West German SPAS) from the shuttle's payload bay, which satellite, once launched, took the first full-view photos of the shuttle in space; and the first person to fly a second STS mission—Robert Crippen.

Dr Ride's colorful *Challenger* patch bears the names of the STS-7 mission crew. *Challenger* again became Sally Ride's vehicle on 5 October 1984, with STS mission 13, during which she became the first US woman astronaut to fly a second shuttle mission. The STS-13 set other firsts: the first seven-member orbiter crew—Crippen, Garneau, Leetsma, McBride, Ride, Scully-Power and Sullivan; the first EVA by a US woman—Kathryn Sullivan; the launching of the Earth Radiation Budget Satellite; the first Canadian in space—Marc Garneau; the first oceanographer to conduct oceanographic studies from space—Paul Scully-Power; and a demonstration of satellite refueling. On 18 August 1986, Dr Sally K Ride was named Special Assistant for Strategic Planning, responsible for reviewing NASA's near-to long-term strategies.

**Above: EASE is an acronym for
'Experimental Assembly of Structures in
EVA.' During STS 61-B, featuring the orbiter**
Atlantis, **astronauts Jerry Ross** *(left)* **and
Sherwood Spring experiment with
component construction for a permanent
manned space station, slated at the time to
become operational in the 1990s. Besides
twice building and disassembling a 45-foot
truss tower composed of 93 six and one-half
and four and one-half foot beams, they built
and disassembled a triangular component
(shown here), designed by MIT, nine times.**

**In their 12 hours of EVA, the astronauts
proved the feasibility of building structures
up to four stories in space, even without the
use of the self-contained Manned
Maneuvering Unit (MMU), designed to
enhance astronaut EVA mobility. It was thus
conjectured that with use of MMU,
structures such as the planned permanent
space station could be built with relative
ease. Ross and Spring were accompanied in**

**the seven-day mission by crewmates Cleave,
Neri, O'Connor, Shaw and Walker.**

**Right: Astronaut Bruce McCandless II is
shown here during his historical EVA on 7
February 1984, which was the first use of the
nitrogen-propelled, hand-controlled Manned
Maneuvering Unit, which allows astronauts
far greater EVA mobility than the previously-
used restrictive tethers. Designed by Martin
Marietta, the MMU weighs 240 lbs, and
utilizes one and one-half pound thrusters, fed
from dual propellant tanks. The left hand
control is for direction of flight and the right
hand control takes care of roll, pitch and
yaw.**

**Shortly after this photo was taken, Robert
L Stewart tried out the MMU that
McCandless is using here, and two days later,
both McCandless and Stewart tested another,
similar unit. Their** *Challenger* **STS 41-B
mission crewmates were Brand, Gibson and
McNair.**

186

Above: In contrast to the all-white tank used on the first two Space Shuttle orbiter flights, the huge external fuel tank on all subsequent flights was painted rust red, in a weight-saving move that trimmed the craft by three-eighths of a ton!

Challenger mission STS-9 was also designated 41-A, instituting a new STS classification system—the 4 stands for the NASA fiscal year of the mission (1984); the 1 stands for the mission's launch site (1 for the Kennedy Space Center, and 2 for the Vandenburg AFB); and the A stands for the mission's intended place in the year's mission schedule (before B and C, etc). The new designation allows the mission itself to be documented even if it falls behind schedule and takes place after a mission that was originally scheduled to take place later than it (for example, STS 51-F took place after STS 51-G). In addition, each orbiter vehicle has its designation: OV-101, *Enterprise*; OV-102, *Columbia*; OV-99, *Challenger*; OV-103, *Discovery*; and OV-104, *Atlantis*.

Right: At 11:38 am on Tuesday, 28 January 1986, *Challenger* STS 51-L lifted off in what looked like a perfect launch. Just 73 seconds later, what had been the shared jubilation of schoolchildren, crew families, mission personnel and millions of other Americans became a horrible explosion, intense confusion and heated arguments as to whether the orbiter program should be allowed to continue.

The solid fuel rockets blasting out of control across the sky like gigantic Fourth of July fireworks, the ball of mockingly beautiful flame which mimed the deceptive, perfect initial liftoff, and the debris that floated down for an hour after the explosion sealed with finality the truth that citizen observer/payload specialist Christa McAuliffe would never teach her classroom lesson from space, and that her six crewmates would never again walk down the ramp from a space vehicle in eager anticipation of rejoining their families and friends.

The failure of an 'O-ring' seal in the solid fuel rocket on *Challenger*'s co-pilot's side of the shuttle allowed a fierce tongue of flame to escape the rocket's side near the external fuel tank and, within seconds, the rocket slammed into and ruptured the fuel tank, causing the liquid hydrogen and liquid oxygen fuels contained within the tank to mix rapidly, which resulted in the devastating explosion that terminated orbiter mission 51-L 73.6 seconds into its flight.

The 14 shuttle missions that were to have followed *Challenger* 51-L in 1986 were indefinitely postponed. On 2 October 1986, NASA gave the go-ahead for the testing of redesigned solid fuel rockets for the STS orbiters, under conditions that simulated engine stresses similar to those of the 51-L accident. After much soul-searching and a thorough investigation of the tragedy, NASA is currently planning to resume shuttle flights in February of 1988.

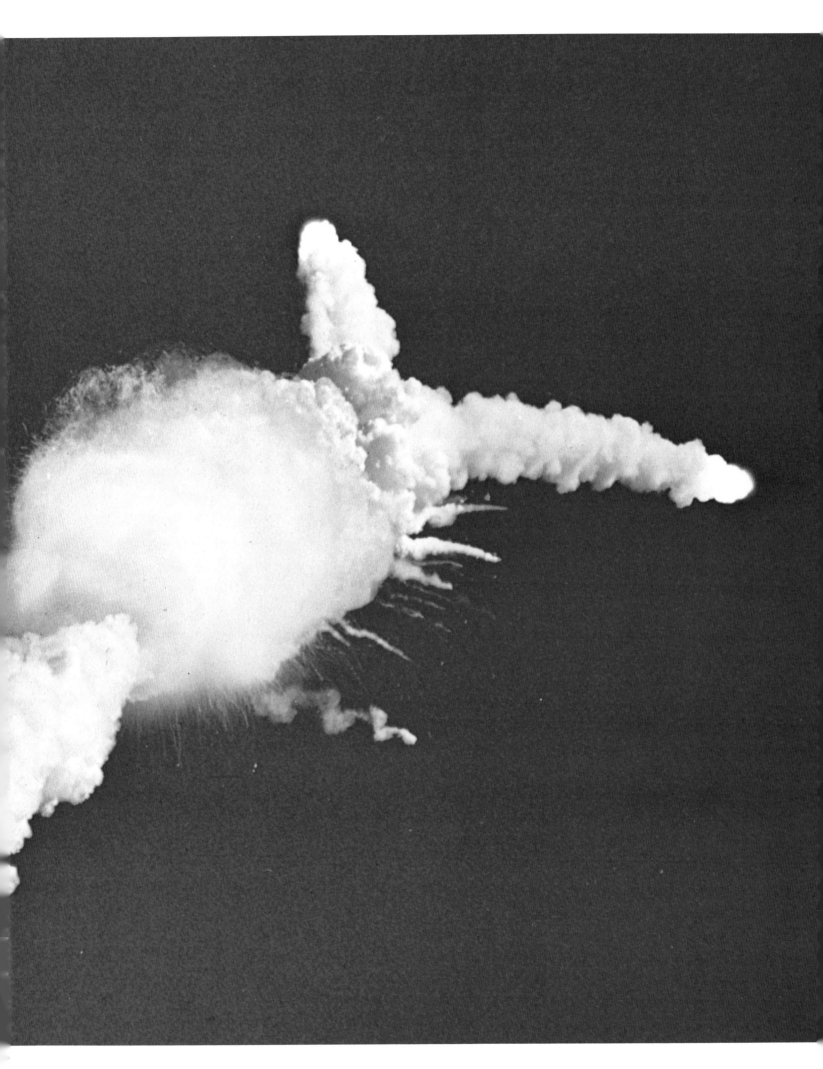

Index

Above: The crewmembers of *Challenger* STS 51-L are shown here as they walked out of the Operations and Checkout Building at the Kennedy Space Center on the cold crisp morning of 28 January 1986. They were Francis R Scobee, Judith A Resnik, Ronald E McNair, Michael J Smith, Christa McAuliffe, Ellison Onizuka, and Gregory B Jarvis.

Overleaf: The face of Concord, New Hampshire high school teacher Sharon Christa McAuliffe, here soon to board the 51-L spacecraft, *Challenger,* shown behind her. She was 37 years old and one of millions who, nineteen years before, found themselves sorrowing over the tragic death of the Apollo 1 astronauts.

Acknowledgments and Credits

The authors and publisher would like to thank the following people who have helped in the preparation of this book: Richard Glassman, who designed it; Thomas G. Aylesworth, who edited it; John K Crowley, who did the photo research; Cynthia Klein, who prepared the index; Timothy Jacobs, who wrote the additional captions for the 1987 edition; Bill Yenne, who designed the additional pages for the 1987 edition and Pamela Berkman, who revised the index for the 1987 edition.

All pictures were supplied by The National Aeronautics and Space Administration, with the following exceptions:
Brown Regional—Planetary Data Center 36-37, 38, 39 (bottom), 41, 42, 43, 44 (bottom), 45, 46 (top), 47, 48, 49, 50, 51
FPG: 22 (bottom), 24 (bottom)
Wide World Photos: 188-189